CHEMISTRY 112

Lecture Guided Workbook

**UNIVERSITY
OF
SOUTH CAROLINA**

Department of Chemistry and Biochemistry

D. L. FREEMAN

D. REGER

S. GOODE

A. TAYLOR-PERRY

**QDE PRESS
2019**

Copyright © 2020 by Freeman, Reger, Goode, Taylor-Perry
Copyright © 2020 Illustrations by QDE Press Inc.

All rights reserved.

Permission in writing must be obtained from the publisher before any part of this work may be reproduced or transmitted in any form or by any means, electronic or mechanical, including photocopying and recording, or by any information storage or retrieval system.

Printed in the United States of America.

ISBN: 978-1-938535-20-8

QDE Press Inc.
Montgomery, Al 36117
www.qdepress.com

Table of Contents

Lecture Notes	**Slide numbers**
Chapter 12 _ Solutions	1-71
Chapter 13_ Chemical Kinetics	72-154
Chapter 14_ Chemical Equilibrium	155-222
Chapter 15_ Solutions of Acids and Bases	223-299
Chapter 16_ Reactions of Acids and Bases	300-379
Chapter 17_ Chemical Thermodynamics	380-436
Chapter 18_ Electrochemistry	437-490
Periodic Table	Inside Back Cover

Lecture Notes

Chapter 12

Solutions

12.1 Solution Concentration

- **Objectives**
  - Express the concentration of a solution using several different units
  - Convert between different concentration units

Solution Concentration

- Concentrations are all fractions.
  - The numerator contains the quantity of solute.
  - The denominator contains the quantity of either solution or solvent.

Solution Concentration

- There are a number of ways to express concentration. You have seen:
  - molarity
  - mole fraction
  - mass percentage

Previously Used Concentration Units

Molarity (Chapter 4):

$$M = \frac{\text{moles of solute}}{\text{liter of solution}}$$

Mole fraction (Chapter 6):

$$\chi_A = \frac{\text{mol A}}{\text{mol A} + \text{mol B} + \text{mol C} + \cdots}$$

Previously Used Concentration Units

Similar to percentage composition (Chapter 3):

$$\text{Mass percentage} = \frac{\text{grams solute}}{\text{grams solution}} \times 100\%$$

Example: Mass Percentage Concentration

- Determine the mass percentage concentration of a solution prepared by dissolving 3.00 g of NaCl (58.44 g/mol) in 150 g of H_2O.

Molality

- Molality (*m* or molal) is defined as

$$\text{molality} = \frac{\text{moles of solute}}{\text{kilogram of solvent}}$$

Example: Molality Calculation

- Determine the molality of a solution prepared by dissolving 3.00 g NaCl (58.44 g/mol) in 150 g of H_2O.

Test Your Skill: Molality

- How many grams of NaCl (58.44 g/mol) must be added to 500. g of water to prepare a 0.75m NaCl solution?

Concentration Units

Concentration Unit	Numerator Units (solute)	Denominator Units
Mass %	Grams	100 g solution
Molarity	Moles	1 L solution
Molality	Moles	1 kg solvent
Mole Fraction	Moles	Total moles of solution

Example: Concentration Conversion

- Express the concentration of a 3.00% H_2O_2 solution as
  - molality
  - mole fraction

Test Your Skill: Molality

- Calculate the molality and mole fraction of ethanol (C_2H_5OH; 46.07 g/mol) in a glass of wine that has an alcohol concentration of 12.5 mass percent.

Example: Conversion to Molarity

- Converting most concentration units to molarity usually involves using the density of the solution to convert mass to units of volume.

- Calculate the molarity of a 12.0% sulfuric acid solution (H_2SO_4; 98.08 g/mol) having a density of 1.080 g/mL.

12.2 Principles of Solubility

- **Objectives**
  - Define solubility and describe how to determine if a solution is saturated, unsaturated, or supersaturated.
  - Describe the energetics of the solution process.
  - Predict the relative solubilities of substances in different solvents based on solute-solvent interactions.

Principles of Solubility

- For most substances, there is a limit to the quantity of solute that dissolves in a specific quantity of solvent.

- A dynamic equilibrium exists between the solute particles in a solution and the undissolved solute.

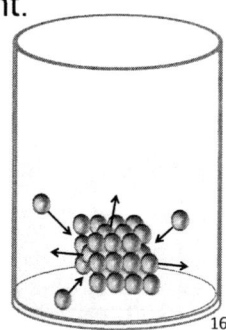

Definitions

- **Solubility** is the concentration of solute that exists in equilibrium with excess of that substance.

- A **saturated solution** has a concentration of solute equal to its solubility.

Definitions

- An **unsaturated solution** is one that has a solute concentration less than the solubility.

- A **supersaturated solution** is one with a solute concentration that is greater than the solubility.
  - Supersaturation is an unstable condition.

Supersaturation

Seed crystal

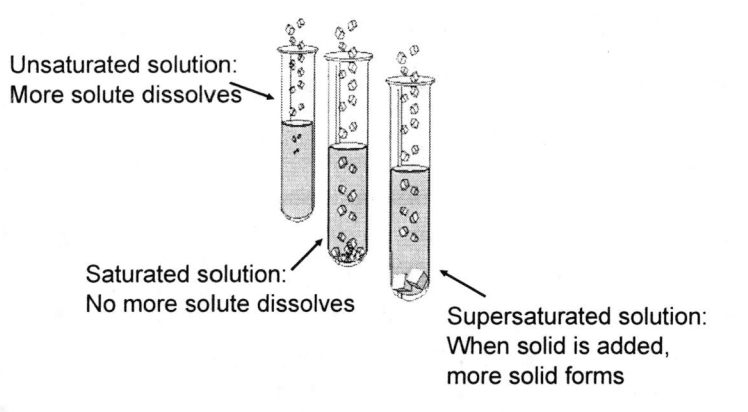

Unsaturated solution: More solute dissolves

Saturated solution: No more solute dissolves

Supersaturated solution: When solid is added, more solid forms

Solute-Solvent Energetics

- Most spontaneous processes are exothermic.
- **Enthalpy of solution**: the enthalpy change that accompanies the dissolution of one mole of solute.
- When a solid and a liquid mix to form a solution, the enthalpy change arises mainly from changes in the intermolecular attractions.

The Solution Process

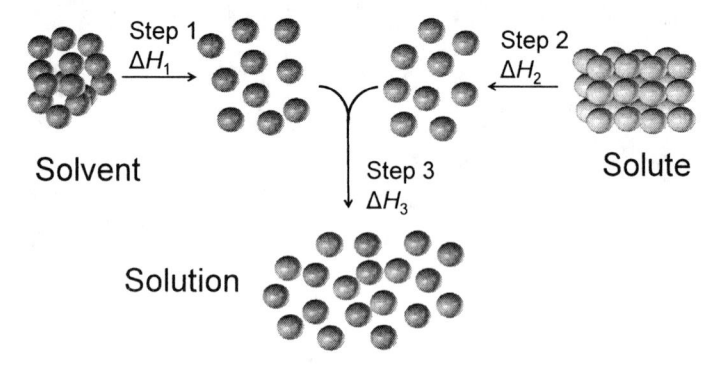

- Steps 1 and 2 are endothermic; Step 3 is exothermic.

Enthalpy of Solution: $\Delta H_{soln} = \Delta H_1 + \Delta H_2 + \Delta H_3$

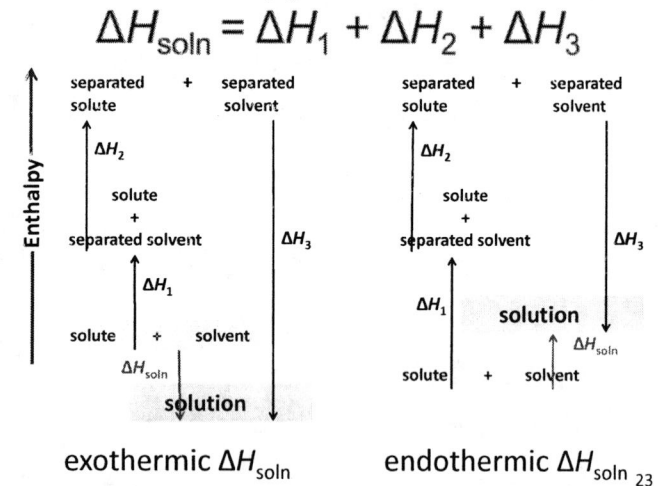

exothermic ΔH_{soln} endothermic ΔH_{soln}

Spontaneity

- Most spontaneous processes are **exothermic**.
  - Some endothermic processes are spontaneous.
- Experiments show that a second factor, an **increase in disorder**, also favors spontaneous change.

Spontaneous Mixing of Gases

- An example of increasing disorder as a driving force is illustrated by the mixing of gases.

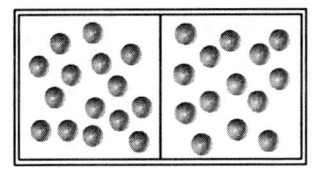

Separated gases   Spontaneously mixed

Disorder and Spontaneity

- An increase in disorder generally accompanies the mixing of molecules in the formation of a solution.
  - Ammonium nitrate is very soluble in water even though the process is quite endothermic (ΔH = +28 kJ/mol). The driving force is disorder.

Solubility of Molecular Compounds

- Relative solubilities can often be predicted by comparing the relative types of interactions between solute, solvent, and solute-solvent.
  - London dispersion, dipole-dipole, and hydrogen bonding
- When the solute and solvent have similar interactions, the solubility is high.
  - "Like dissolves like"

Example: Relative Solubility

(a) Is iodine (I_2) more soluble in water or in hexane (C_6H_{14})?

(b) Is methanol (CH_3OH) more soluble in water or in hexane (C_6H_{14})?

Ionic Compounds Dissolving in Water

- Many ionic compounds are soluble in water.
- Soluble ionic compounds separate into ions in solution.
- Electrostatic forces between the ions and water molecules favor the solubility process.
- Ionic compounds are seldom soluble in non-polar liquids.

Interaction of Ions in Water

- The interaction of ions with water molecules is called hydration.

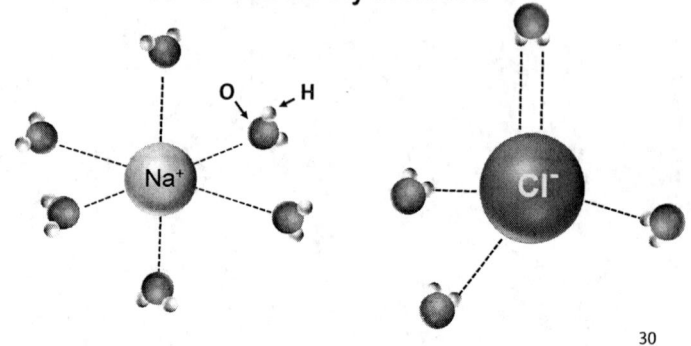

Ionic Compounds in Water

- When an ionic compound dissolves in water, disorder changes.
  - Separating the ions increases disorder.
  - Separating the water molecules increases disorder.
  - Hydrating the ions, which restricts some water molecules, decreases disorder.
- When the balance of enthalpy and disorder is favorable, an ionic compound dissolves.

Example: Relative Solubility

Is potassium chloride (KCl) more soluble in water or in hexane (C_6H_{14})?

12.3 Effects of Pressure and Temperature on Solubility

- **Objectives**
  - State the effects of pressure and temperature on solubility.
  - Calculate the solubility of gases using Henry's law.
  - Explain why changes in pressure do not appreciably change the solubilities of solids and liquids.
  - Relate the sign of the enthalpy of solution to the increase or decrease of solubility with temperature.

Pressure and Solubility

- Pressure has very little effect on the solubilities of liquids and solids.
- The solubility of gases in a liquid depends on the pressure of the gas.
- **Henry's Law**: The solubility of a gas is directly proportional to its partial pressure at any given temperature:
$$C = kP$$

Henry's Law Constants in Water for Various Gases (molal/atm)

Gas	0 °C	20 °C	40 °C	60 °C
CO_2	7.60 x 10^{-2}	3.91 x 10^{-2}	2.44 x 10^{-2}	1.63 x 10^{-2}
N_2	1.03 x 10^{-3}	7.34 x 10^{-4}	5.55 x 10^{-4}	4.85 x 10^{-4}
O_2	2.21 x 10^{-3}	1.43 x 10^{-3}	1.02 x 10^{-3}	8.71 x 10^{-4}

Example: Henry's Law Calculation

- Water at 20 °C is saturated with air that contains CO_2 at a partial pressure of 8.0 torr. What is the molal concentration of CO_2 in the solution?

Solubility and Temperature

- Experiments show that the way solubility changes with temperature depends on the sign of the enthalpy of solution.
  - Solubility increases with increasing temperature if ΔH_{soln} is positive (endothermic).
  - Solubility decreases with increasing temperature if ΔH_{soln} is negative (exothermic).

Temperature Dependence on Solubility

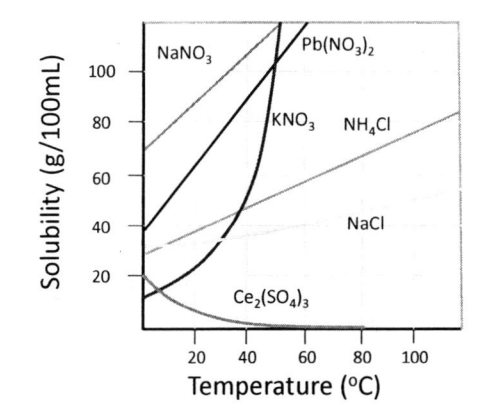

12.4 Colligative Properties of Solutions

- **Objectives**
  - List and define the colligative properties of solutions.
  - Relate the values of colligative properties to the concentrations of solutions.
  - Calculate the molar masses of solutes from measurements of colligative properties.

Colligative Properties of Solutions

- **Colligative property**: Any property of a solution that changes in proportion to the concentration of solute particles.
  - For non-electrolyte solutes, the identity of the particle has no influence on colligative property.
- Many colligative properties are directly related to the lowering of solvent vapor pressure by the presence of solute particles.

Effect of Solute on Vapor Pressure

- The vapor pressure of a solvent in a solution is lower than that of the pure solvent.

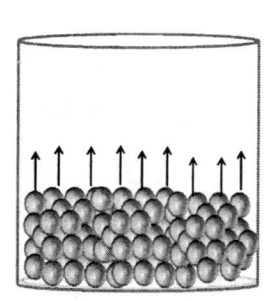

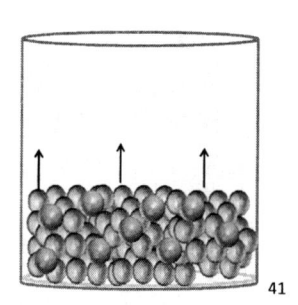

Raoult's Law

- **Raoult's law**: The vapor pressure of solvent above a dilute solution equals the mole fraction of the solvent times the vapor pressure of the pure solvent.
  - $P_{solv} = \chi_{solv} P^o_{solv}$
- Another form of this equation gives the lowering of the vapor pressure.
  - $\Delta P_{solv} = \chi_{solute} P^o_{solv}$

Example: Raoult's Law

- At 27°C, the vapor pressure of benzene is 104 torr. What is the vapor pressure of a solution that has 0.100 mol of naphthalene dissolved in 1.90 mol of benzene?

Boiling Point Elevation

- Because a solute lowers the vapor pressure of the solvent, it raises the boiling point of the solution.
- The normal boiling point is the temperature at which the vapor pressure equals 1 atm.

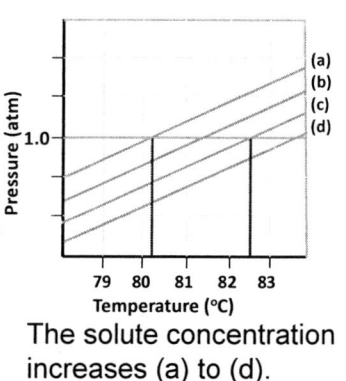

The solute concentration increases (a) to (d).

Boiling Point Elevation

- The relationship between change in boiling point and solute concentration:
  - $\Delta T_b = mk_b$ where m is the molal concentration of solute and k_b is the boiling point constant for the solvent

Solvent	B.P. (°C)	k_b (°C/m)
Acetic acid	117.90	3.07
Benzene	80.10	2.53
Water	100.00	0.512

Freezing Point Depression

- Solute particles interfere with the ability of solvent particles to form a crystal and freeze. Thus, it takes a lower temperature to freeze solvent from a solution than from the pure solvent. This lowering is freezing point depression.

Freezing Point Depression

- The relationship between change in freezing point and solute concentration:
  - $\Delta T_f = mk_f$ where m is the molal concentration of solute and k_f is the freezing point constant for the solvent.

Solvent	F.P. (°C)	k_f (°C/m)
Acetic acid	16.60	3.90
Benzene	5.51	4.90
Water	0.00	1.86

Example: Calculate k_f

- Benzophenone freezes at 48.1 °C. Adding 1.05 grams of urea, $(NH_2)_2CO$ (60.06 g/mol), to 30.0 g of benzophenone forms a solution that freezes at 42.4 °C. Calculate k_f for benzophenone.

Test Your Skill: Freezing Point Depression

- Benzophenone freezes at 48.1 °C and has a k_f of 9.8 °C/m. A 2.50 g sample of solute with a molar mass of 130.0 g/mol is dissolved in 32.0 g of benzophenone. Calculate the freezing point of the solution.

Osmosis

- **Osmosis** is the diffusion of a liquid through a semipermeable membrane.
  - Semipermeable membranes allow small molecules like water to pass through them.

Osmosis

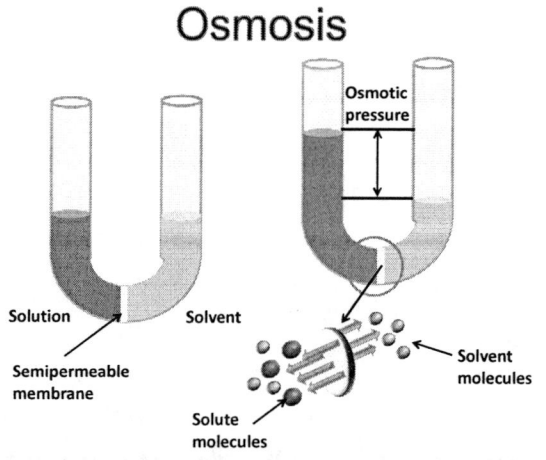

- The increased level of liquid produces an additional pressure, called **osmotic pressure**.

Osmotic Pressure

- Osmotic pressure is a colligative property, and can be calculated by the equation:

$$\Pi = MRT$$

where:

Π = osmotic pressure

M = molar concentration of solute

R = 0.082 L·atm/mol·K

T = temperature in kelvins

Example: Osmotic Pressure Problem

- A 5.70 mg sample of protein is dissolved in water to give 1.00 mL of solution. Calculate the molar mass of the protein if the solution has an osmotic pressure of 6.52 torr at 20 °C.

Test Your Skill: Osmotic Pressure

- Osmotic pressure experiments yield a concentration of 0.0021 M for a solution made by dissolving 9.4 mg of a peptide and diluting to 2.00 mL. Calculate the molar mass of the peptide.

Summary of Colligative Properties

Property	Symbol	Concentration Unit	Constant
Decrease in vapor pressure	ΔP	Mole fraction	P^o
Boiling point elevation	ΔT_b	Molal	k_b
Freezing point depression	ΔT_f	Molal	k_f
Osmotic pressure	Π	Molar	RT

12.5 Colligative Properties of Electrolyte Solutions

- **Objectives**
  - Predict the ideal van't Hoff factor of ionic solutes
  - Calculate the colligative properties for solutions of electrolytes

Electrolyte Solutions

- The colligative properties of electrolyte solutions are more pronounced because electrolytes separate into ions in solution.
- The **van't Hoff factor**, i, is the number of ions formed when a formula unit of an ionic compound dissolves.
- Non-electrolyte solutions have a van't Hoff factor of 1.

The van't Hoff Factor

- In dilute solution, the van't Hoff factor for salts approaches the number of ions produced by one formula unit of the substance.
  - NaCl → Na$^+$(aq) + Cl$^-$(aq) $i=2$
  - MgBr$_2$ → Mg^{2+}(aq) + 2 Br$^-$(aq) $i=3$
- The van't Hoff factor generally decreases as the concentration increases.

Accounting for the van't Hoff Factor

- The equations for calculating colligative properties of non-electrolyte solutes can be modified to account for electrolytes.
- $\Delta T_b = imk_b$
- $\Delta T_f = imk_f$
- $\Pi = iMRT$

Example: van't Hoff Factor

- Arrange the following aqueous solutions in order of increasing boiling points: 0.03 m urea (a non-electrolyte), 0.01 m NaOH, 0.02 m BaCl$_2$, 0.01 m Fe(NO$_3$)$_3$.

Example: Freezing Point of an Electrolyte Solution

- Water freezes at 0.0 °C and has a k_f of 1.86 °C/m. A 3.50 g sample of NaCl is dissolved in 50.0 g of water. Calculate the freezing point of the solution.

12.6 Mixtures of Volatile Substances

- **Objectives**
  - Calculate the vapor pressure of each component and the total vapor pressure over an ideal solution.
  - Explain how fractional distillation works.

Mixtures of Volatile Substances

- In a solution of two or more volatile compounds, all components of the mixture are in equilibrium with their vapors.
- An **ideal solution** is one in which all volatile components obey Raoult's law for all compositions.
- $P_A = X_A P°_A$, $P_B = X_B P°_B$ ... etc.

Mixtures of Toluene and Benzene Form Nearly Ideal Solutions

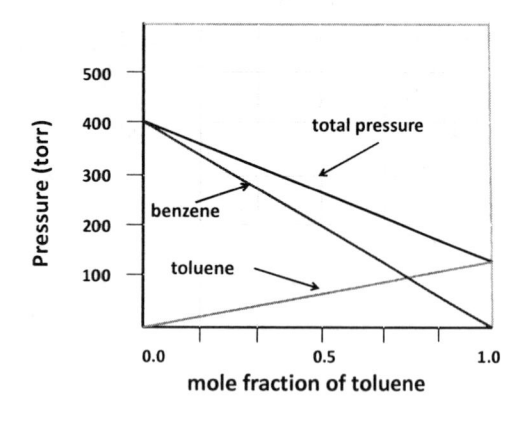

Example: Vapor Pressure of Solutions

- At 27°C, the vapor pressure of carbon tetrachloride (CCl_4) is 127 torr and that of chloroform ($CHCl_3$) is 212 torr. What is the partial pressure of each substance, and the total vapor pressure above a solution that contains 0.40 mol of CCl_4 and 0.60 mol of $CHCl_3$?

Test Your Skill: Vapor Pressure of Solutions

- From the previous example, calculate the mole fraction of $CHCl_3$ **in the vapor**.

Distillation

- **Distillation** is the separation of a mixture of components based on differences in volatility (vapor pressure) by repeated evaporation and condensation of the mixture.
- The vapor always contains a larger mole fraction of the more volatile component as seen in the previous "Test Your Skill."

Distillation Apparatus

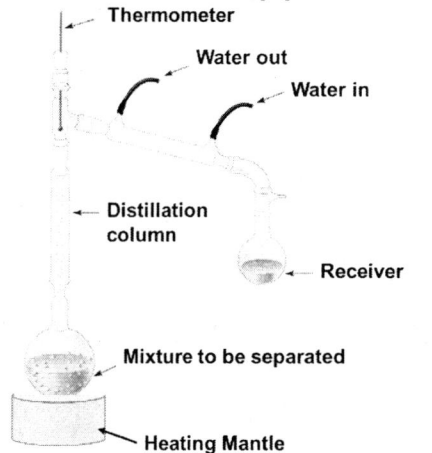

- Thermometer
- Water out
- Water in
- Distillation column
- Receiver
- Mixture to be separated
- Heating Mantle

Deviations from Raoult's Law

Some liquid-liquid solutions deviate from Raoult's law.

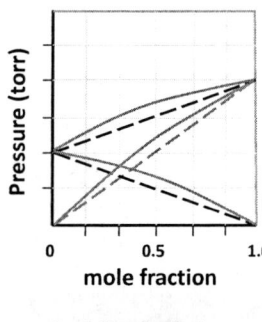

Positive deviation

Negative deviation

Large Deviation from Ideality

- Deviations sometimes are large enough to produce a maximum or minimum in the vapor pressure curve.

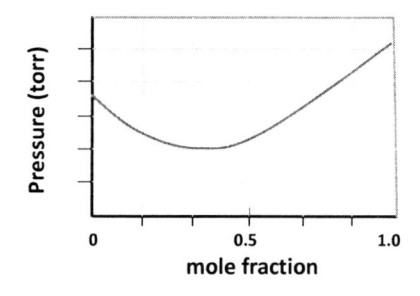

Azeotropes

- The composition of the solution at minimum or maximum on the vapor pressure curve is a constant-boiling mixture called an **azeotrope**.

- Any mixture of compounds that forms an azeotrope cannot be separated from the azeotropic composition by distillation.

Chapter 13

Chemical Kinetics

13.1 The Rates of Reactions

- **Objectives**
  - Relate the changes in concentration over time to the rate of reaction.
  - Calculate the instantaneous rate of reaction from experimental data.
  - Use stoichiometry to relate the rate of reaction to changes in the concentrations of reactants and products.

Kinetics

- **Kinetics** is the study of the rates of chemical reactions.
  - Rate is defined as the change of concentration (c) per unit time (t):

$$rate = \frac{\Delta c}{\Delta t}$$

Rates of Chemical Reactions

- Reaction rates are always positive.
- Rate is expressed as either the appearance of the product (+) or disappearance of the reactant (−) per unit time.

$$rate = \frac{\Delta c}{\Delta t} = \frac{\Delta [products]}{\Delta t} = \frac{-\Delta [reactants]}{\Delta t}$$

- Rates can be expressed in several ways:
  - M/s or mol/(L·s)
  - $M \cdot s^{-1}$ or $mol \cdot L^{-1} \cdot s^{-1}$

Rate of Reaction

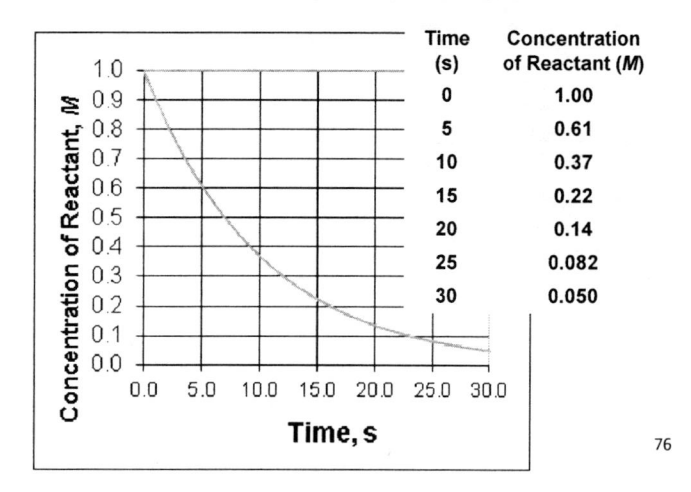

Time (s)	Concentration of Reactant (M)
0	1.00
5	0.61
10	0.37
15	0.22
20	0.14
25	0.082
30	0.050

Average Rates

- When a rate is measured over a time interval, the result is an average rate.
- Average rates are not very useful, because they depend on specific intervals.

Instantaneous Reaction Rate

- The instantaneous rate of the reaction is equal to the slope of the line drawn tangent to the curve at time t.
- Rate at 10 s = $\dfrac{0.72 - 0 \; M}{20 - 0 \; s}$ = 0.036 M/s

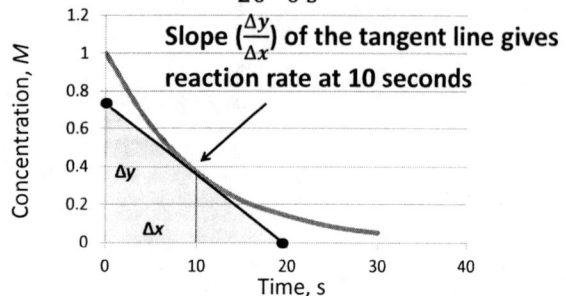

Slope ($\dfrac{\Delta y}{\Delta x}$) of the tangent line gives reaction rate at 10 seconds

Reaction Rate and Stoichiometry

- The relative rates of disappearance of reactants or appearance of products depend on the reaction stoichiometry.
  - $2 \text{ HBr (g)} \rightarrow H_2 \text{ (g)} + Br_2 \text{ (g)}$
- Two moles of HBr are consumed for every one mole of H_2 or Br_2 that is formed. So the rate of change of [HBr] is double that of either $[H_2]$ or $[Br_2]$.
  - $-\dfrac{1}{2} \times \dfrac{\Delta[\text{HBr}]}{\Delta t} = \dfrac{\Delta[H_2]}{\Delta t} = \dfrac{\Delta[Br_2]}{\Delta t}$

Rate and Reaction Stoichiometry

- For any reaction:
  - $aA + bB \rightarrow cC + dD$
  - The reaction rate is given by:
  - $rate = -\dfrac{1}{a} \times \dfrac{\Delta[A]}{\Delta t} = -\dfrac{1}{b} \times \dfrac{\Delta[B]}{\Delta t}$
  - $= \dfrac{1}{c} \times \dfrac{\Delta[C]}{\Delta t} = \dfrac{1}{d} \times \dfrac{\Delta[D]}{\Delta t}$
- Note the signs as well as the coefficients.

Example: Rate and Reaction Stoichiometry

- For the reaction
  $5H_2O_2 + 2MnO_4^- + 6H^+ \rightarrow 2Mn^{2+} + 5O_2 + 8H_2O$
  - The experimentally determined rate of disappearance of MnO_4^- is 2.2×10^{-3} M/s.
  - Calculate the reaction rate.
  - What is the rate of appearance of O_2?

13.2 Relationships Between Rate and Concentration

- **Objectives**
  - Define a rate law to express the dependence of the rate of reaction on the concentrations of the reactants.
  - Determine the reaction order from the rate law.
  - Use initial concentrations and initial rates of reactions to determine the rate law and rate constant.

Relating Rate and Concentration

- **Rate law**: dependence of rate on concentration.
  - Analysis of many experiments shows that the rate of a reaction is proportional to the product of the concentrations of the reactants raised to some power.
  - For the reaction $aA + bB \rightarrow$ products the rate law is the given by:
  $$rate = k[A]^x[B]^y$$

Differential Rate Law

- rate=$k[A]^x[B]^y$
  - x and y are the orders of the reaction in [A] and [B] respectively.
  - The overall order of the reaction is $x+y$.
  - x and y are usually small integers, but may be zero, negative or fractions.
  - k is the specific rate constant.

Rate and Concentration

- The reaction orders are determined by noting the effect of changing the concentration of each reactant on the rate.
- Given the initial concentrations and initial rates you can determine the order with respect to each reactant and calculate the overall order of the reaction.
- The specific rate constant, k, is evaluated once the orders in the rate law are known.

Dependence of Rate on Order

Reactant → Products

	[R]	Relative Rate
First order rate law rate=$k[R]^1$	1 2 3	1 2 3
Second order rate law rate=$k[R]^2$	1 2 3	1 4 9
Zero order rate law rate=$k[R]^0$	1 2 3	1 1 1

Initial Rate Method: A + B → Products

- Experiments are performed in which initial concentrations of [A] and [B] are individually varied.
- The rate is measured at the instant A and B are mixed.
- The time interval of measurements is small so that the measured rate is approximately equal to the instantaneous rate.

Example: Initial Rate

- Use initial rate data to determine the rate law for

$$F_2 + 2 ClO_2 \rightarrow 2 FClO_2$$

Experiment	Initial conc. [F_2], M	Initial conc. [ClO_2], M	Initial Rate M/s
1	0.10	0.010	1.2x10^{-3}
2	0.10	0.040	4.8x10^{-3}
3	0.20	0.010	2.4x10^{-3}

Test Your Skill: Initial Rate

- Given the following data write the rate law for the reaction:

$$2NO + 2H_2 \rightarrow N_2 + 2H_2O$$

Trial	Initial [NO]	Initial [H_2]	Initial Rate M/s
1	0.00570	0.140	7.01x10^{-5}
2	0.00570	0.280	1.40x10^{-4}
3	0.0114	0.140	2.81x10^{-4}

13.3 Dependence of Concentrations on Time

- **Objectives**
  - Evaluate concentration-time behaviors to write a rate law.
  - Contrast differential and integrated forms of the rate law.
  - Calculate the concentration-time behavior for a first order reaction from the rate law and the rate constant

13.3 Dependence of Concentrations on Time

- **Objectives continued**
  - Relate half-life and rate constant and calculate concentration-time behavior from the half life of a first order reaction
  - Calculate the concentration-time behavior for a second order reaction from the rate law and the rate constant

Two Forms of Rate Laws

Differential Rate Law
- describes the dependence of rate on concentration
- rate = $k[R]^x$
- $\frac{-\Delta[R]}{\Delta t} = k[R]^x$

Integrated Rate Law
- describes the dependence of concentration on time
- $[R] = f(t)$

Zero Order Rate Law

- The reaction rate is independent of the reactant concentration [R] for a zero order reaction.

$$R \rightarrow \text{products}$$

$$\text{rate} = \frac{-\Delta[R]}{\Delta t} = k[R]^0 \quad \text{(differential)}$$

$$\frac{-\Delta[R]}{\Delta t} = k$$

$$[R] = [R]_o - kt \quad \text{(integrated)}$$

Zero-Order Rate Laws

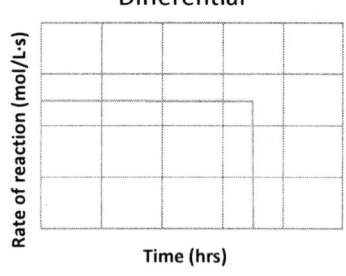

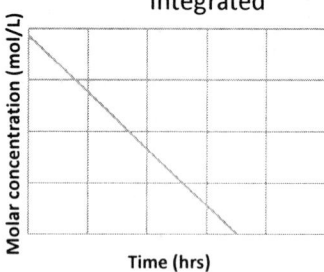

Differential

[Graph: Rate of reaction (mol/L·s) vs. Time (hrs) — horizontal line]

A plot of rate vs. time yields a straight line with a slope of zero. Rate is constant.
rate = $k[R]^0 = k$

Integrated

[Graph: Molar concentration (mol/L) vs. Time (hrs) — straight line with negative slope]

A plot of concentration vs. time yields a straight line with slope = $-k$
$[R] = [R]_o - kt$

First Order Rate Law

The reaction rate depends on concentration raised to the first power.

$$R \rightarrow products$$

Differential rate law
- rate = $k[R]$
- $\dfrac{-\Delta[R]}{\Delta t} = k[R]$
- $\dfrac{\Delta[R]}{[R]} = -k\Delta t$

Integrated rate law
- $[R] = [R]_o e^{-kt}$

  or
- $\ln[R] = \ln[R]_o - kt$

Concentration-Time Dependence

- Integrated first order rate law:
$$[R] = [R]_o e^{-kt}$$

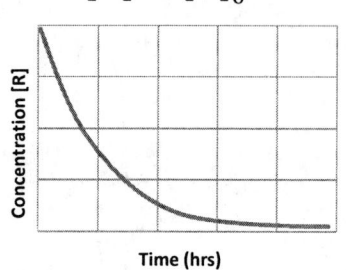

[Graph: Concentration [R] vs. Time (hrs) — exponential decay curve]

A plot of concentration vs. time yields a curve.

Concentration-Time Dependence

- Integrated first order rate law:
$$\ln[R] = \ln[R]_0 - kt$$

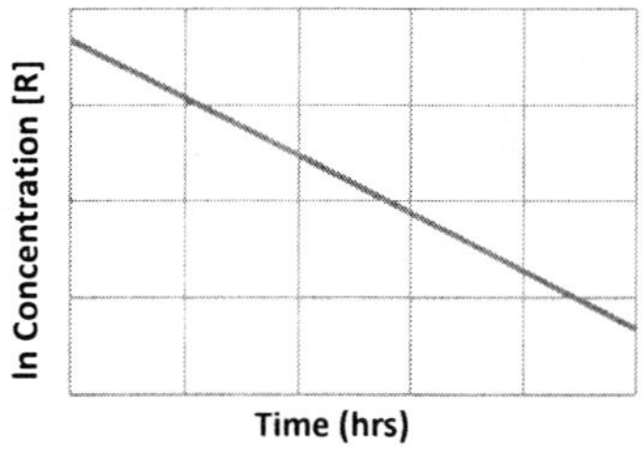

A plot of ln[R] vs. time yields a straight line.

Concentration-Time Dependence

Graph of conc. vs. time → Is graph a straight line? → (Yes) System is zero order

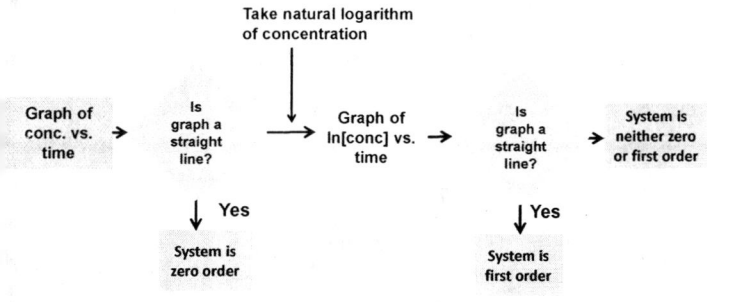

Take natural logarithm of concentration → Graph of ln[conc] vs. time → Is graph a straight line? → (Yes) System is first order

Otherwise: System is neither zero or first order

Example: First Order Rate Law

- $C_{12}H_{22}O_{11} + H_2O \rightarrow C_6H_{12}O_6 + C_6H_{12}O_6$
  sucrose + water → glucose + fructose
  - Experiments show the reaction is first order, $k = 6.2 \times 10^{-5}$ s^{-1}.

- If $[C_{12}H_{22}O_{11}]_0 = 0.40\ M$, what is $[C_{12}H_{22}O_{11}]$ after 2 hours?

Test Your Skill: First Order Reaction

- How long did it take for the sucrose concentration in the same experiment to drop from 0.40 M to 0.30 M?

  $k = 6.2 \times 10^{-5}$ s^{-1}

Half-Life

- **Half-life, $t_{1/2}$**, is the time required for the initial concentration to decrease by ½.

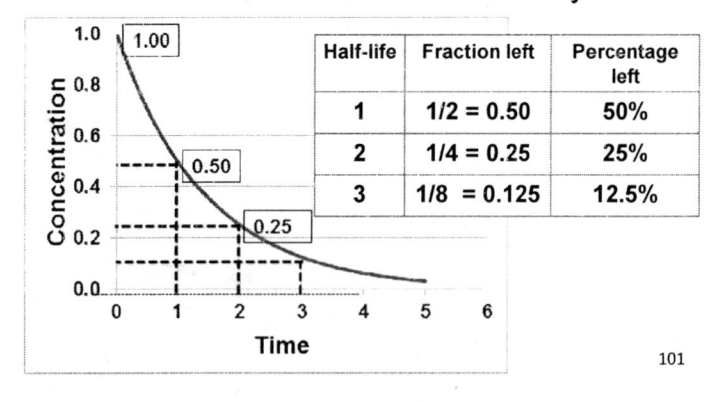

Half-life	Fraction left	Percentage left
1	1/2 = 0.50	50%
2	1/4 = 0.25	25%
3	1/8 = 0.125	12.5%

First Order Half-Life

- For a first order reaction,
$$\ln[R] = \ln[R]_o - kt$$
- When $t = t_{1/2}$, $[R] = ½[R]_o$.
- Substitute and rearrange to solve for $t_{1/2}$,
$$t_{1/2} = \frac{0.693}{k}$$
- The half-life of a first order reaction is independent of the concentration.

Calculating Half-Life

- $k = 6.2 \times 10^{-5}$ s^{-1} for the reaction
- $C_{12}H_{22}O_{11} + H_2O \rightarrow C_6H_{12}O_6 + C_6H_{12}O_6$

- Calculate the half-life.

Radiocarbon Dating

- The age of objects that were once living can be found by ^{14}C dating, because the concentration of ^{14}C is constant in the biosphere.
- When an organism dies, the ^{14}C content decreases with first order kinetics ($t_{1/2}$ = 5730 years).
- Scientists calculate the age of an object from the concentration of ^{14}C in a sample.

Example: ^{14}C Dating

- A sample of wood has 58% of the ^{14}C originally present. Calculate the age of the wood sample. The half-life of ^{14}C is 5730 years.

Test Your Skill: ^{14}C Dating

- The ^{14}C concentration in an archeological artifact has decayed by 22.0% of its original concentration. Calculate the age of the artifact. The half-life of ^{14}C is 5730 years.

Second Order Rate Law

The reaction rate depends on concentration raised to the second power.

$$R \rightarrow products$$

Differential rate law
- rate = $k[R]^2$
- $\dfrac{-\Delta[R]}{\Delta t} = k[R]^2$
- $\dfrac{\Delta[R]}{[R]^2} = -k\Delta t$

Integrated rate law
- $\dfrac{1}{[R]} = \dfrac{1}{[R]_o} + kt$

Concentration-Time Dependence

- Integrated second order rate law:
$$\dfrac{1}{[R]} = \dfrac{1}{[R]_o} + kt$$

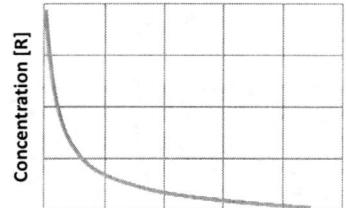

A plot of concentration vs. time yields a curve.

Concentration-Time Dependence
- Integrated second order rate law:
$$\frac{1}{[R]} = \frac{1}{[R]_o} + kt$$

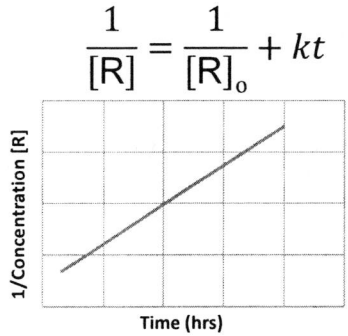

A plot of 1/[R] vs. time yields a straight line.

Example: Second Order Rate Law

- The reaction
$$2\ NOCl \rightarrow 2NO + Cl_2$$
obeys the rate law
rate = 0.020 $M^{-1}s^{-1}$ [NOCl]^2
- If the initial concentration of NOCl is 0.050 M, calculate the concentration after 30 minutes.

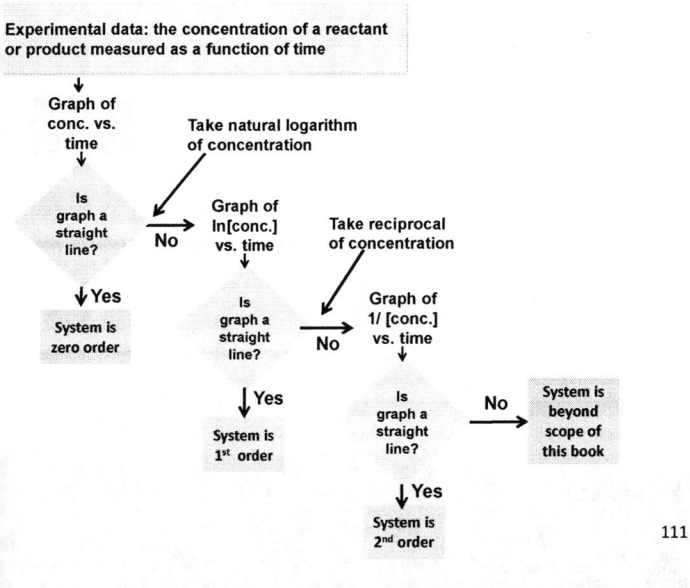

Example: Order of Reaction

- Given the experimental data for the decomposition of 1,3-pentadiene shown below, determine the order of the reaction.

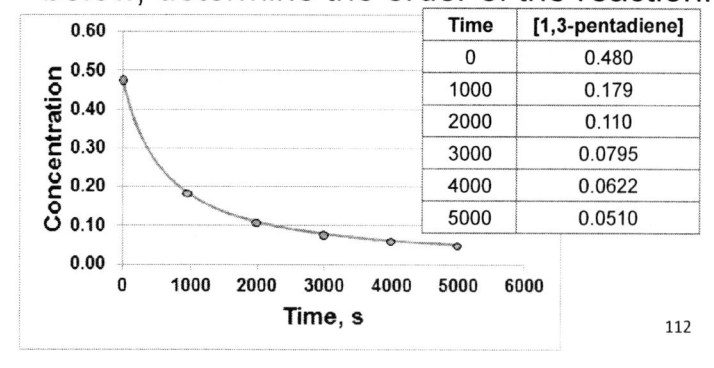

Time	[1,3-pentadiene]
0	0.480
1000	0.179
2000	0.110
3000	0.0795
4000	0.0622
5000	0.0510

Example: Order of Reaction cont.

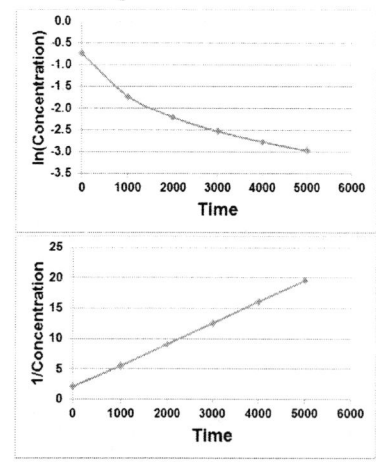

A plot of ln[1,3-pentadiene] vs. time

A plot of 1/[1,3-pentadiene] vs. time gives a straight line where the slope is equal to k.

Summary: R → products

	Order		
	Zero	First	Second
Differential rate law	rate = k	rate = $k[R]$	rate = $k[R]^2$
Integrated rate law	$[R] = [R]_o - kt$	$\ln[R] = \ln[R]_o - kt$	$1/[R] = 1/[R]_o - kt$
Units of k	$M\,s^{-1}$	s^{-1}	$M^{-1}\,s^{-1}$

Test Your Skill: Integrated Rate Laws

- The reaction $C_4H_8 \rightarrow 2C_2H_4$ obeys the rate law
  rate = $87s^{-1} [C_4H_8]$
- How long will it take for 95% of the C_4H_8 to decompose?

13.4 Mechanisms I: Macroscopic Effects – Temperature and Energetics

- **Objectives**
  - Describe the effect of changing temperature on the rate of reaction.
  - Use the collision theory to relate collision frequency, activation energy, and steric factor to the rate of reaction.
  - Relate temperature, activation energy, and rate constant through the Arrhenius equation.

Influence of Temperature on k

Reactions proceed at faster rates at higher temperatures.

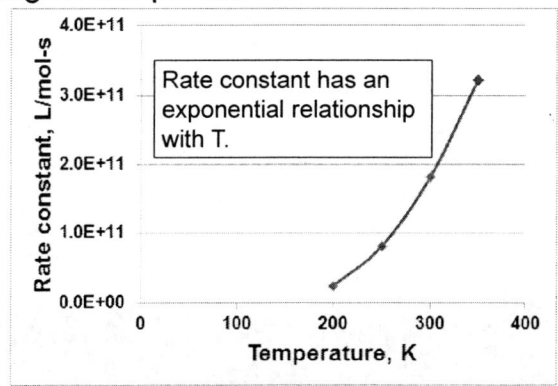

Rate constant has an exponential relationship with T.

Collision Frequency

- The reaction rate is proportional to the collision frequency, Z, the number of molecular collisions per second.
- For the reaction A + B → Products
- $Z = Z_o[A][B]$ represented in general as $Z = Z_o[\text{colliding species}]$
  - Z_o is a proportionality constant.

Collision Theory

- Experiments show that not all collisions result in the formation of product.
- **Activation energy (E_a):** the minimum collision energy required for reaction to occur.

The Activated Complex

Activated complex: the least-stable (highest energy) arrangement of atoms that occurs in the reaction.

NO + O_3 → [activated complex] → NO_2 + O_2

[NO-O_3]*

NO + O_3

$\updownarrow E_a$

NO_2 + O_2

Energy

Reaction Coordinate

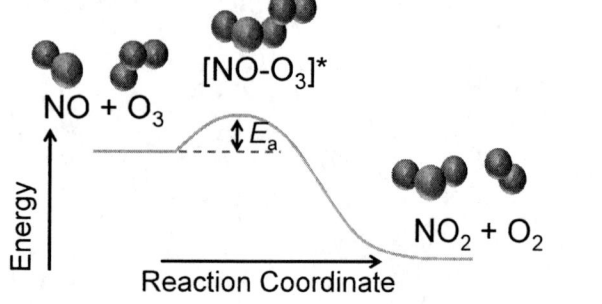

Influence of Temperature on Kinetic Energy

- The fraction of collisions with energy in excess of E_a equals $e^{-E_a/RT}$.
- The rate of reaction is proportional to the collision frequency times the fraction of collisions with energy in excess of E_a.

Orientation of Reactants

$NO + O_3 \rightarrow [ON\text{-}O_3]^* \rightarrow NO_2 + O_2$

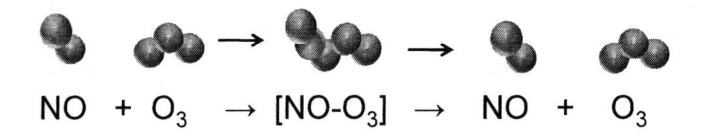

$NO + O_3 \rightarrow [NO\text{-}O_3] \rightarrow NO + O_3$

The Steric Factor

- The steric factor, p, accounts for orientation of reactants.
- The rate of reaction is equal to the steric factor times the collision frequency times the fraction of collisions with energy in excess of E_a:
- rate = $p \times Z_o[\text{colliding species}] \times e^{-E_a/RT}$

The Arrhenius Equation

- rate = $p \times Z_o$ [colliding species] $\times e^{-E_a/RT}$
- Combine p and Z_o into a term A to simplify:

  rate = $A \times$ [colliding species] $\times e^{-E_a/RT}$

- Measurements to determine the rate law show that

  rate = $k \times$ [colliding species] so

- $k = Ae^{-E_a/RT}$
  - The rate constant is exponentially dependent on temperature.

Evaluating Activation Energy

- Take the natural log of both sides of the Arrhenius equation:

$$k = Ae^{-E_a/RT}$$

$$\ln k = \ln A - E_a/R \times \frac{1}{T}$$

A plot of $\ln k$ vs. $1/T$ gives a straight line with a slope of $-E_a/R$ and an intercept of $\ln A$.

Example: Measuring Activation Energy

- Determine E_a for the reaction

  $2NO_2 \rightarrow 2NO + O_2$

  given the following data:

k (M⁻¹s⁻¹)	T (K)	ln k	1/T (K⁻¹)
0.003	500		
0.291	600		
7.39	700		

Measurements at Two Different Temperatures

- Only two points are needed to calculate a slope resulting in the following equation for activation energy:

$$\ln\left(\frac{k_1}{k_2}\right) = \frac{-Ea}{R}\left(\frac{1}{T_1} - \frac{1}{T_2}\right)$$

- Rates can be used in place of rate constants.

Test Your Skill: Arrhenius Equation

- The rate constant of a reaction exactly doubles when the temperature is changed from 25 °C to 36 °C. Calculate the activation energy for this reaction.

13.5 Catalysis

- **Objectives**
  - Define catalysis and identify heterogeneous, homogeneous, and enzymatic catalysts.
  - Draw energy-level diagrams for catalyzed and uncatalyzed reactions.

Catalysis

- A **catalyst** is a substance that increases the reaction rate but is not consumed in the reaction.
- A catalyst provides an alternate reaction path with a lower activation energy.

Homogeneous Catalysis

- A **homogeneous catalyst** is one that is present in the same phase as the reactants.
- Bromide ion catalyzes the decomposition of aqueous hydrogen peroxide.

$$2H_2O_2(aq) \xrightarrow{Br^-(aq)} 2H_2O(\ell) + O_2(g)$$

Heterogeneous Catalysis

- A **heterogeneous catalyst** is one that is present in a different phase from the reactants.
- The decomposition of aqueous hydrogen peroxide can also be catalyzed by solid MnO_2.

$$2H_2O_2(aq) \xrightarrow{MnO_2(s)} 2H_2O(\ell) + O_2(g)$$

Enzyme Catalysis

- **Enzymes** are large molecules (macromolecules) that catalyze specific biochemical reactions.
- Enzymes can increase the rates of reactions by factors as large as 10^{14}.
- Enzymes are specific for the reactions they catalyze.
- Enzymes are active under mild reaction conditions.

13.6 Mechanisms II: Microscopic Effects – Collisions between Molecules

- **Objectives**
  - Describe a chemical reaction as a sequence of elementary processes.
  - Write the rate law from an elementary step and determine its molecularity.
  - Predict the experimental rate law from the mechanism and differentiate among possible reaction mechanisms by examining experimental rate data.

Reaction Mechanisms

- An **elementary step** is a chemical equation that describes an actual molecular-level collision.
- A **mechanism** is a sequence of elementary steps that leads from reactants to products.

Elementary Steps

- The gas phase reaction
$$2NO + O_2 \rightarrow 2NO_2$$
is believed to occur by the following sequence of elementary steps:

$2NO \rightarrow N_2O_2$	Step 1
$N_2O_2 + O_2 \rightarrow 2NO_2$	Step 2
$2NO + O_2 \rightarrow 2NO_2$	overall reaction

- The elementary steps sum to the overall chemical equation.

Reaction Intermediates

- An **intermediate** is a substance produced in an early step and consumed in a later step.

$2NO \rightarrow N_2O_2$
$N_2O_2 + O_2 \rightarrow 2NO_2$
$2NO + O_2 \rightarrow 2NO_2$

- N_2O_2 is the intermediate.

Identifying Catalysts and Intermediates

- Catalysts appear first on the reactant side.
- Intermediates appear first on the product side.
- Neither is found in the overall chemical equation.

Example: Reaction Mechanisms
- Write the overall chemical equation and identify any catalyst and intermediates in the following mechanism.

Step 1:
$$H_2O_2(aq) + 2Br^-(aq) \rightarrow Br_2(aq) + 2OH^-(aq)$$
Step 2:
$$H_2O_2(aq) + Br_2(aq) + 2OH^-(aq) \rightarrow 2Br^-(aq) + 2H_2O(\ell) + O_2(g)$$

Molecularity
- The **molecularity** of an elementary step is the number of reactant species involved in that step.
- Most elementary steps are either **unimolecular** (involving a single molecule) or **bimolecular** (involving two molecules).

Molecularity
- The gas phase reaction
$$2NO + O_2 \rightarrow 2NO_2$$
is believed to occur in two bimolecular steps:

$2NO \rightarrow N_2O_2$	bimolecular
$N_2O_2 + O_2 \rightarrow 2NO_2$	bimolecular
$2NO + O_2 \rightarrow 2NO_2$	overall reaction

Rate Laws for Elementary Reactions

- The rate of an elementary step is proportional to the concentration of each reactant species raised to the power of its coefficient in the equation:
- Step 1: $2NO \rightarrow N_2O_2$
    $rate_1 = k_1[NO]^2$
- Step 2: $N_2O_2 + O_2 \rightarrow 2 NO_2$
    $rate_2 = k_2[N_2O_2][O_2]$

Test Your Skill: Elementary Steps

- Identify the molecularity and write the rate law for the elementary step:

$$H_2 + Cl \rightarrow H_2Cl$$

Rate-Limiting Steps

- The overall rate of a multistep reaction is determined by its slowest step, called the rate-limiting step.
- The fast steps that follow the rate-limiting step have no effect on the overall rate law.
- The fast steps that precede the rate-limiting step usually affect the concentrations of the reactant species in the rate-determining step.

Multistep Reaction Mechanisms

- Consider the two step reaction
$$2NO + O_2 \rightarrow 2NO_2$$
1: $2NO \rightarrow N_2O_2$ $rate_1 = k_1[NO]2$
2: $N_2O_2 + O_2 \rightarrow 2NO_2$ $rate_2 = k_2[N_2O_2][O_2]$

Determination of Rate Laws

- If the first step is rate limiting, the rate law is:

$rate_1 = k_1[NO]2$

- If the first step is rapid and the second step is rate limiting, the rate law is:
$rate_2 = k_2[N_2O_2][O_2]$
  - Rate laws written in terms of intermediates are generally not useful, because the concentration of an intermediate is difficult to measure.

Complex Reaction Mechanisms

- The concentration of the intermediate can often be expressed in terms of the concentrations of the reactants from which it is formed.

    Step 1: $2NO \rightarrow N_2O_2$
    Step 2: $N_2O_2 + O_2 \rightarrow 2NO_2$

Step 1: $2NO \rightleftharpoons N_2O_2$

Typically a fast first step reaches equilibrium.

$rate_1 = k_1[NO]^2$ Forward Step 1

$rate_{-1} = k_{-1}[N_2O_2]$ Reverse Step 1

Because the rates at equilibrium are equal:

$rate_1 = rate_{-1}$

$k_1[NO]^2 = k_{-1}[N_2O_2]$

$[N_2O_2] = \dfrac{k_1}{k_{-1}}[NO]^2$

Step 2: $N_2O_2 + O_2 \rightarrow 2NO_2$

$rate_2 = k_2[N_2O_2][O_2]$

- Substituting the expression for $[N_2O_2]$ from the previous slide into $rate_2$:

  $rate_2 = k_2 \dfrac{k_1}{k_{-1}} [NO]^2[O_2]$

- Combine all the rate constants and write the rate law:

  $rate = k[NO]^2[O_2]$

- The reaction is second order in NO and first order in O_2.

Test Your Skill: Mechanisms

- For the reaction
  $$2NO_2 + O_3 \rightarrow N_2O_5 + O_2$$
  the experimentally determined rate law is $rate = k[NO_2][O_3]$.
- Identify the rate limiting step in the proposed two-step mechanism:

  $NO_2 + O_3 \rightarrow NO_3 + O_2$ Step 1
  $NO_3 + NO_2 \rightarrow N_2O_5$ Step 2

The Hydrogen-Iodine Reaction

- $H_2 + I_2 \rightarrow 2HI$ rate = $k[H_2][I_2]$
  - For many years this reaction was believed to occur as a single bimolecular step.
- From more recent data, a very different mechanism is likely.

  $I_2 \rightleftharpoons 2I$ Fast, reversible
  $I + H_2 \rightleftharpoons H_2I$ Fast, reversible
  $H_2I + I \rightarrow 2HI$ Slow

- Both mechanisms give the same rate law.

Enzyme Catalysis

- Most enzymes follow the **Michaelis-Menten** mechanism.
- Step 1: the enzyme binds the substrate in a rapid, reversible reaction to form a complex.

$$E + S \underset{k_{-1}}{\overset{k_1}{\rightleftharpoons}} ES$$

This step reaches equilibrium.

Enzyme Catalysis

- Step 2: the product forms from the complex, and the enzyme is released.

$$ES \xrightarrow{k_2} E + P$$

This step is the rate-limiting step.

- rate = $k_2[ES]$

Enzyme Catalysis

- Under normal conditions, [S] is much greater than $[E]_o$.
- Nearly all the enzyme is bound to the substrate, so $[ES] \sim [E]_o$, and
    rate = $k_2[E]_o$ = constant
- The reaction rate is zero order in substrate, S.

Chapter 14

Chemical Equilibrium

14.1 The Equilibrium Constant

- **Objectives**
  - Describe equilibrium systems and write the equilibrium constant expression for any chemical reaction.
  - Evaluate the equilibrium constant from experimental data.

14.1 The Equilibrium Constant

- **Objectives continued**
  - Relate the expression for the equilibrium constant to the form of the balanced equation.
  - Convert between equilibrium constants in which the concentrations of gases are expressed in different units.

Chemical Equilibrium

- **Chemical equilibrium**: a state in which the tendency of the reactants to form products is balanced by the tendency of the products to form reactants.

$$\text{reactant} \rightleftharpoons \text{products}$$

- At equilibrium, the reaction has not stopped; the forward and reverse reaction rates are the same.

Equilibrium Systems

- Phase change:

$$H_2O(\ell) \rightleftharpoons H_2O(g)$$

  - At equilibrium, liquid water evaporates at the same rate the water vapor condenses.

- Chemical equilibrium:

$$N_2(g) + 3H_2(g) \rightleftharpoons 2NH_3(g)$$

  - At equilibrium, the rate at which ammonia forms is equal to the rate at which it decomposes.

Establishing Equilibrium

$$2NO_2 (g) \rightleftharpoons N_2O_4 (g)$$

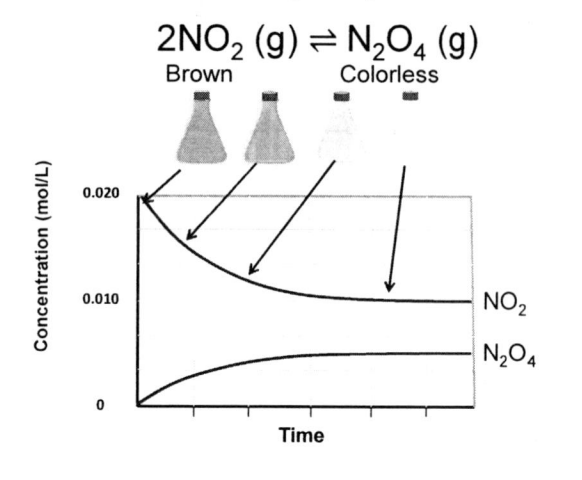

Brown   Colorless

$$2NO_2 (g) \rightleftharpoons N_2O_4 (g)$$

Starting with NO_2 Starting with N_2O_4

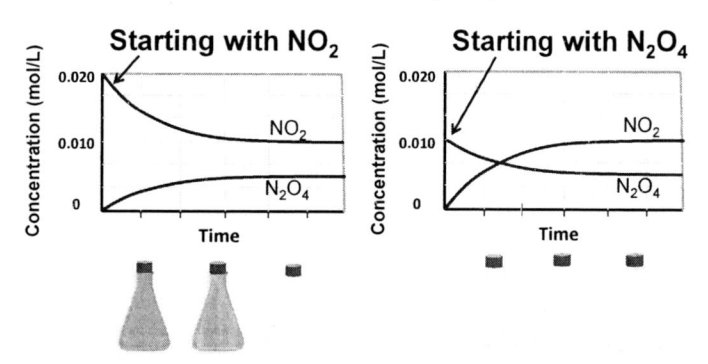

Initial and Equilibrium Concentrations

$2NO_2(g) \rightleftharpoons N_2O_4(g)$ Temp = 317 K

Initial Conc., *M*		Equil. Conc., *M*	
[NO_2]	[N_2O_4]	[NO_2]	[N_2O_4]
2.00×10^{-2}	0.00	1.03×10^{-2}	4.86×10^{-3}
0.00	1.00×10^{-2}	1.03×10^{-2}	4.86×10^{-3}
3.00×10^{-2}	1.00×10^{-2}	1.85×10^{-2}	1.57×10^{-2}
4.00×10^{-2}	0.00	1.61×10^{-2}	1.19×10^{-2}

Square brackets indicate molar concentrations.

Experimental Equilibrium Constants

$$2NO_2(g) \rightleftharpoons N_2O_4(g) \qquad K_{eq} = \frac{[N_2O_4]}{[NO_2]^2}$$

Initial Conc., M		Equil. Conc., M		$K_{eq} = \frac{[N_2O_4]}{[NO_2]^2}$
[NO$_2$]	[N$_2$O$_4$]	[NO$_2$]	[N$_2$O$_4$]	
2.00x10^{-2}	0.00	1.03x10^{-2}	4.86x10^{-3}	
0.00	1.00x10^{-2}	1.03x10^{-2}	4.86x10^{-3}	
3.00x10^{-2}	1.00x10^{-2}	1.85x10^{-2}	1.57x10^{-2}	
4.00x10^{-2}	0.00	1.61x10^{-2}	1.19x10^{-2}	

The Law of Mass Action

- For a general chemical reaction

$$aA + bB \rightleftharpoons cC + dD$$

- The equilibrium constant is given by

$$K_{eq} = \frac{[C]^c[D]^d}{[A]^a[B]^b}$$

Example: Expression for K_{eq}

- Write the expression for the equilibrium constant for each reaction below.
- $2O_3(g) \rightleftharpoons 3O_2(g)$

- $CO(g) + H_2O(g) \rightleftharpoons H_2(g) + CO_2(g)$

Example: Determining K_{eq} from Experiment

- Consider the following reaction:

$$2H_2(g) + S_2(g) \rightleftharpoons 2H_2S(g)$$

Experiment shows that at equilibrium, there are 2.50 mol H_2, 1.35×10^{-5} mol S_2, and 8.70 mol H_2S in a 12.0 L container.

- Calculate K_{eq}.

K_{eq} and the Chemical Equation

- The form of the equilibrium expression depends on the balanced equation.

Reaction 1: $H_2(g) + \frac{1}{2} O_2(g) \rightleftharpoons H_2O(g)$

$$K_1 = \frac{[H_2O]}{[H_2][O_2]^{1/2}}$$

Reaction 2: $2H_2(g) + O_2(g) \rightleftharpoons 2H_2O(g)$

$$K_2 = \frac{[H_2O]^2}{[H_2]^2[O_2]} = K_1^2$$

K_{eq} and the Chemical Equation

Reaction 2: $2H_2(g) + O_2(g) \rightleftharpoons 2H_2O(g)$

$$K_2 = \frac{[H_2O]^2}{[H_2]^2[O_2]}$$

Reaction 3: $2H_2O(g) \rightleftharpoons 2H_2(g) + O_2(g)$

$$K_3 = \frac{[H_2]^2[O_2]}{[H_2O]^2} = \frac{1}{K_2}$$

- For any reaction; $K_{rev} = \dfrac{1}{K_{forward}}$

Test Your Skill: K_{eq}

- $N_2O_4(g) \rightleftharpoons 2NO_2(g)$ $\quad K_{eq} = 4.6 \times 10^{-3}$

- Determine K_{eq} for the reaction
 $2NO_2(g) \rightleftharpoons N_2O_4(g)$

- Determine K_{eq} for the reaction
 $NO_2(g) \rightleftharpoons \frac{1}{2}N_2O_4(g)$

Adding Chemical Equations

- When chemical equations are added to generate a new equation, the equilibrium constant of the new equation is equal to the product of the equilibrium constants.

$$2NO_2(g) \rightleftharpoons N_2O_4(g) \quad K_1$$
$$N_2O_4(g) + O_2(g) \rightleftharpoons 2NO_3(g) \quad K_2$$
$$\overline{2NO_2(g) + O_2(g) \rightleftharpoons 2NO_3(g) \quad K_3 = K_1 \times K_2}$$

Concentration and Pressure

$$N_2O_4(g) \rightleftharpoons 2NO_2(g)$$

- K_c indicates molar concentrations (M) are used.

$$K_C = \frac{[NO_2]^2}{[N_2O_4]}$$

- K_P indicates pressure (atm).

$$K_P = \frac{P_{NO_2}^2}{P_{N_2O_4}}$$

Relating K_P and K_c

$$K_P = \frac{P_{NO_2}^2}{P_{N_2O_4}}$$ For any gas $P = \frac{n}{V}RT$.

Because $\frac{n}{V} = [\text{Conc}]$

$$K_P = \frac{([NO_2]RT)^2}{[N_2O_4]RT} = K_c RT$$

- In general $K_P = K_c(RT)^{\Delta n}$, where

Δn = moles gaseous product – moles gaseous reactant.
- When $\Delta n = 0$, $K_P = K_c$.

Example: Converting K_P and K_c

- Calculate K_P for each of the following reactions.
1. K_c is 5.0×10^6 at 700 K for
    $$2SO_2(g) + O_2(g) \rightleftharpoons 2SO_3(g)$$
2. K_c is 6.0×10^{34} at 300 K for
    $$NO(g) + O_3(g) \rightleftharpoons NO_2(g) + O_2(g)$$

Test Your Skill: K_P and K_c

- Calculate K_c for

$$PCl_5(g) \rightleftharpoons PCl_3(g) + Cl_2(g)$$

Given $K_P = 229$ at 425 °C.

14.2 The Reaction Quotient

- **Objectives**
  - Write the expression for Q, the reaction quotient, and contrast with the expression for K_{eq}.
  - Compare Q and K_{eq} to determine the direction in which a reaction proceeds when a system is not at equilibrium.

The Reaction Quotient

- **Reaction Quotient, Q**, has the same algebraic form as K_{eq}, but is evaluated with current concentrations, rather than equilibrium concentrations.

    aA + bB ⇌ cC + dD

$$Q = \frac{[C]^c[D]^d}{[A]^a[B]^b}$$

Determining Direction of Reaction

- $Q < K_{eq}$: ratio of products to reactants is too small, reaction will move in forward direction to reach equilibrium.
- $Q = K_{eq}$: system is at equilibrium.
- $Q > K_{eq}$: ratio of products to reactants is too large, reaction will move in reverse direction to reach equilibrium.

Determining Direction of Reaction

- Place the symbols for Q and K_{eq} on a number line.
- The reaction will proceed in the direction that moves Q toward K_{eq}.

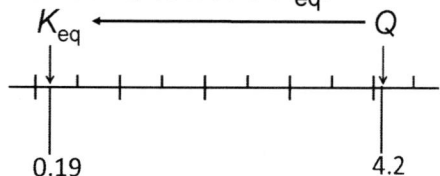

K_{eq} ← ——— Q

0.19   4.2

Determining Direction of Reaction

$$2NO_2(g) \rightleftharpoons N_2O_4(g) \quad K_c = \frac{[N_2O_4]}{[NO_2]^2} = 45.8$$

Initial Conc., M		Equil. Conc., M		Q, direction
[NO$_2$]	[N$_2$O$_4$]	[NO$_2$]	[N$_2$O$_4$]	
2.00x10^{-2}	0.00	1.03x10^{-2}	4.86x10^{-3}	0, forward
0.00	1.00x10^{-2}	1.03x10^{-2}	4.86x10^{-3}	
3.00x10^{-2}	1.00x10^{-2}	1.85x10^{-2}	1.57x10^{-2}	

Test Your Skill: Direction of Reaction

$CH_4(g) + H_2O(g) \rightleftharpoons CO(g) + 3H_2(g) \quad K_c = 5.67$

Initial concentrations:

$CH_4(g)$ + $H_2O(g)$ $\rightleftharpoons$ $CO(g)$ + $3H_2(g)$
0.100 M   0.200 M   0.500 M   0.800 M

Calculate the reaction quotient, Q, and determine the direction the reaction proceeds to reach equilibrium.

14.3 Le Chatelier's Principle

- **Objectives**
  - Predict the response of an equilibrium system to changes in conditions by applying Le Chatelier's principle.
  - Determine how changes in temperature influence the equilibrium system.

Le Chatelier's Principle

- Any change to a chemical reaction at equilibrium causes the reaction to proceed in the direction that reduces the effect of the change.

Changing Concentration of a Reactant or Product

- Adding a reactant or product causes the reaction to move in the direction that consumes the added substance.
- Removing a reactant or product causes the reaction to move in the direction that produces the missing substance.

Changing Partial Pressure of a Reactant or Product

$$2SO_3(g) \rightleftharpoons 2SO_2(g) + O_2(g)$$

- An increase in SO_2 partial pressure causes the formation of SO_3 and the consumption of O_2.
- The reaction proceeds in reverse.

Increasing Pressure by Adding an Inert Gas

- Increasing the pressure of a system by adding an inert gas does not change the concentration or partial pressure of the reactants or products.
- Materials that do not react have no influence on the equilibrium.

The Effect of Volume Change

- Increasing the volume causes the reaction to favor the side with the larger number of moles of gas.
- Decreasing the volume causes the reaction to favor the side with the smaller number of moles of gas.
- If the chemical equation has the same number of moles of gas on both sides, there is no effect from changing volume.

Example: Volume Change and Reaction Direction

$$2SO_3(g) \rightleftharpoons 2SO_2(g) + O_2(g)$$

- Increasing the volume causes the reaction to _____.

- Decreasing the volume causes the reaction to _____.

Changes in Temperature

- Changing temperature does not influence concentrations.
- Changing temperature influences K_{eq}.
- We can use Le Chatelier's principle to determine if K_{eq} will get larger or smaller.

Changes in Temperature

- Heat is a product in an exothermic reaction and a reactant in an endothermic reaction.

$$CO(g) + 2H_2(g) \rightleftharpoons CH_3OH(g) \quad \Delta H = -18 \text{ kJ}$$

- The reaction is exothermic (heat is a product). Increasing the temperature causes the reaction to move in reverse.
- The value of K_{eq} will get smaller.

Adding a Catalyst

- A catalyst increases the reaction rate but does not affect the equilibrium (final) concentrations and does not affect the value of K_{eq}.

Test Your Skill: Le Chatelier's Principle

- Consider the reaction

$N_2(g) + 3H_2(g) \rightleftharpoons 2NH_3(g) \quad \Delta H° = -92$ kJ

In which direction will the reaction proceed in response to the following changes?

a. Decreasing the partial pressure of H_2
b. Increasing the concentration of NH_3
c. Increasing the temperature
d. Decreasing the volume
e. Increasing the total pressure by adding argon

14.4 Equilibrium Calculations

- **Objectives**
  - Use a systematic method, the iCe table, to solve chemical equilibria.
  - Calculate equilibrium constants from experimental data and stoichiometric relationships.
  - Calculate the equilibrium concentrations of species in a chemical reaction.

Classes of Equilibrium Calculations

- Equilibrium calculations fall into two categories:
  - K_{eq} can be calculated from measurements of the equilibrium concentrations.
  - Equilibrium concentrations can be calculated from K_{eq} and initial concentrations.

Problem-Solving Strategy

- Write the balanced chemical equation.
- Fill in an "iCe" table.
- Write the expression for the equilibrium constant.
- Substitute concentrations from the iCe table into the equilibrium constant expression.
- Solve the expression.

Example: Determining K_c from Experiment

- A scientist places 1.0 mol of HI in a 10.0 L flask. The equilibrium concentration of I_2 was determined to be 0.020 M. Calculate K_c for:

$$2HI(g) \rightleftharpoons H_2(g) + I_2(g)$$

- The iCe table provides a strategy to solve this problem.

iCe Table

Initial condition: 1.0 mol of HI in a 10.0 L flask.
Equilibrium: [I_2] is 0.020 M.

	2HI(g) ⇌	H_2(g) +	I_2(g)
i (initial, M)	0.10		
C (change, M)			
e (equilibrium, M)			0.020

Given information in bold

The missing concentrations can be determined from given data.

Example: Using K_c and Initial Conc. to Calculate Equilibrium Conc.

- If the initial concentration of CO is 0.028 M and H_2 is 0.14 M, and K_c = 0.50 for

$$CO(g) + H_2(g) \rightleftharpoons CH_2O(g)$$

Calculate the equilibrium concentrations of all components.

Test Your Skill: Equilibrium

- If the initial concentration of PCl_5 is 0.100 M, and K_c = 0.60, calculate the equilibrium concentrations for all components of the following reaction.

$$PCl_5(g) \rightleftharpoons PCl_3(g) + Cl_2(g)$$

Example: Equilibrium

- A researcher studies an industrial process by placing 0.030 mol SO_2Cl_2 in a 100.0 L reactor with 2.0 mol SO_2 and 1.0 mol Cl_2 at 173°C. At this temperature, K_P is 3.0 for $SO_2Cl_2(g) \rightleftharpoons SO_2(g) + Cl_2(g)$.
- Calculate all equilibrium concentrations.

14.5 Heterogeneous Equilibria

- **Objectives**
  - Write equilibrium constant expressions for heterogeneous equilibria.

Heterogeneous Equilibria

- A **heterogeneous system** is one in which the reactants and products are present in more than one phase.

$$CaCO_3(s) \rightleftharpoons CaO(s) + CO_2(g)$$

Heterogeneous Equilibria

$$CaCO_3(s) \rightleftharpoons CaO(s) + CO_2(g)$$

$$K_c' = \frac{[CaO][CO_2]}{[CaCO_3]} \qquad K_P' = \frac{P_{CaO}P_{CO_2}}{P_{CaCO_3}}$$

- The concentration/pressure of a pure solid or pure liquid is constant and is not included in the equilibrium expression.

$$K_c = [CO_2] \qquad K_P = P_{CO_2}$$

Test Your Skill: Heterogeneous Equilibria

$$Hg_2Cl_2(s) \rightleftharpoons 2Hg(\ell) + Cl_2(g)$$

$K_c = $ _____

$$NH_3(g) + HCl(g) \rightleftharpoons NH_4Cl(s)$$

$K_P = $ _____

14.6 Solubility Equilibria

- **Objectives**
  - Write the expression for the solubility product constant.
  - Calculate K_{sp} from experimental data.
  - Calculate solubility of slightly soluble salts from K_{sp}.

Solubility Equilibria

- **Solubility equilibria** reactions are those that involve formation or dissolution of a solid.
  - The net ionic equation for the formation of AgCl is
    $$Ag^+(aq) + Cl^-(aq) \rightleftharpoons AgCl(s)$$
  - The net ionic equation for the dissolution of AgCl is
    $$AgCl(s) \rightleftharpoons Ag^+(aq) + Cl^-(aq)$$

Solubility Product Constant

- For a sparingly soluble solid such as magnesium fluoride,
  $$MgF_2(s) \rightleftharpoons Mg^{2+}(aq) + 2F^-(aq)$$
- K_{sp}, the solubility product constant is:
  $$K_{sp} = [Mg^{2+}][F^-]^2$$

Test Your Skill: Expressions for K_{sp}

Write the chemical equation and the expression for the solubility product constant for:

(a) Iron(III) hydroxide

(b) Lead(II) chloride

Example: K_{sp} Calculation

- A scientist prepares a saturated solution of lead(II) iodide. The equilibrium concentration of Pb^{2+}(aq) is 1.3×10^{-3} M. Calculate K_{sp}.

Test Your Skill: K_{sp}

- In a saturated solution of silver phosphate, experimental measurement shows $[Ag^+] = 1.3 \times 10^{-4}$ M. Calculate K_{sp}.

Solubility

- **Solubility** is the concentration of solute that exists in equilibrium with an excess of that substance.
- The symbol for solubility is a lower case "s."

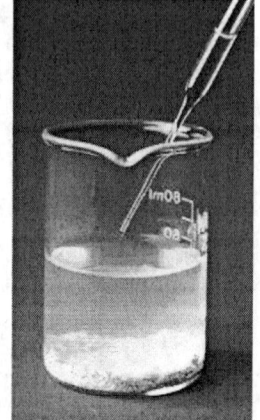

Example: Solubility Calculations

- K_{sp} for silver chloride is 1.8×10^{-10}. Calculate the solubility in mol/L.

Test Your Skill: Solubility

- K_{sp} for calcium hydroxide, $Ca(OH)_2$, is 1.3×10^{-6}. Calculate the molar solubility.

Solubility

- Solubilities of different compounds cannot be predicted by ranking them in order of K_{sp}.

Compound	K_{sp}	Solubility, M
$AgIO_3$	3.1×10^{-8}	1.8×10^{-4}
$Ba(IO_3)_2$	1.5×10^{-9}	7.2×10^{-4}
$La(IO_3)_3$	6.2×10^{-12}	6.9×10^{-4}

14.7 Solubility and the Common Ion Effect

Objectives
- Predict the solubility of a solid in a solution that contains a common ion.
- Use numerical methods to calculate roots of polynomial equations.
- Determine if a precipitate will form under a particular set of conditions.

Common Ions

- Frequently chemists need to calculate the solubility of a sparingly soluble solid in a solution that already contains one of the ions (common ion) that compose the solid.
- For example, when studying the solubility of Ag_2S, if the solution in which we dissolve the Ag_2S already contains Ag^+ or S^{2-}, the solution has ions in common with the solid.

Common Ion Effect

- **Common ion effect**: the effect of adding a solute to a solution that contains a common ion decreases the solubility of the solid.
- Adding NaCl will introduce Cl^- (a common ion) and decrease the solubility of AgCl.

$$AgCl(s) \rightleftharpoons Ag^+(aq) + Cl^-(aq)$$

The effect is consistent with Le Chatelier's principle.

Example: Common Ion Effect

- What is the solubility of $Mg(OH)_2$ in a solution of 0.10 M NaOH?
  K_{sp} for $Mg(OH)_2$ is 8.9 x 10^{-12}.

Example: Common Ion Effect

- Calculate the solubility of $Mg(OH)_2$ in pure water and compare to previous example.

Common Ion Effect

- The solubility of $Mg(OH)_2$
  - Is 1.3 x 10^{-4} M in pure water,
  - Is 8.9 x 10^{-10} M in 0.10 M NaOH.

Test Your Skill: Common Ion Effect
- Calculate the solubility of silver sulfite in 0.100M silver nitrate solution. K_{sp} for silver sulfite is 1.5×10^{-14}.

Forming a Precipitate
- Under conditions when a solubility product reaction proceeds in reverse, a precipitate forms.

$$MX(s) \rightleftharpoons M^+(aq) + X^-(aq)$$

- Calculate Q and compare to K_{sp} to determine if a precipitate will form when two solutions are mixed.

$Q<K_{sp}$	$Q>K_{sp}$
No solid present	Ions form solid

K_{sp} equilibrium

Example: Does a Precipitate Form?
- 25.00 mL of 0.050 M Na_2CO_3 and 10.00 mL of 0.0020 M $Ca(NO_3)_2$ are mixed. Will a precipitate form?
  - According to the solubility rules the likely precipitate is $CaCO_3$.

K_{sp} ($CaCO_3$) = 8.7×10^{-9}

Chapter 15

Solutions of Acids and Bases

Arrhenius Acids and Bases

- **Arrhenius acid**: any substance that produces H$^+$ when dissolved in water.
- **Arrhenius base**: any substance that produces OH$^-$ when dissolved in water.
- Arrhenius acids and bases are limited to aqueous solutions.

15.1 Brønsted-Lowry Acid-Base Systems

- **Objectives**
  - Define Brønsted-Lowry acids and bases.
  - Differentiate between Brønsted-Lowry and Arrhenius acids and bases.
  - Identify conjugate acid-base pairs.
  - Identify the proton transfer in Brønsted-Lowry acid-base reactions.

Brönsted-Lowry Acids and Bases

- **Brönsted-Lowry acid**: proton (H^+) donor
- **Brönsted-Lowry base**: proton (H^+) acceptor
- Brönsted-Lowry acid-base reactions involve the transfer of a proton (H^+) from an acid to a base.
- Brönsted-Lowry acids and bases are not limited to aqueous solutions.

Conjugate Acid-Base Pairs

- **Conjugate acid-base pairs**: molecules or ions that differ by a single proton (H^+).
- The acid contains the proton (H^+) that is transferred to a molecule or ion.
- The conjugate base is the molecule or ion remaining after the loss of the proton (H^+).

Nomenclature in Aqueous Solution

- In aqueous solution, chemists use the following terms interchangeably.
  - Proton
  - $H^+(aq)$ (hydrogen ion)
  - H_3O^+ (hydronium ion)

Conjugate Acid-Base Pairs

H⁺ transferred to the H₂O molecule

$$HF(aq) + H_2O(\ell) \rightleftharpoons F^-(aq) + H_3O^+(aq)$$

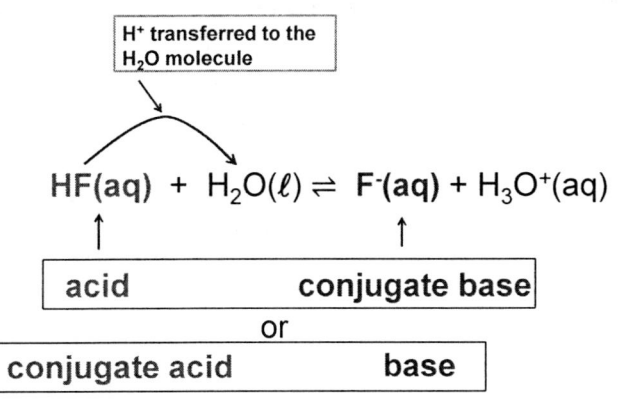

- acid — conjugate base
- or
- conjugate acid — base

Conjugate Acid-Base Pairs
Reverse Reaction

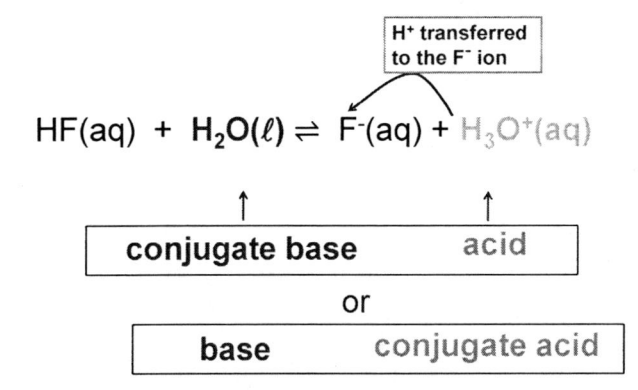

H⁺ transferred to the F⁻ ion

$$HF(aq) + H_2O(\ell) \rightleftharpoons F^-(aq) + H_3O^+(aq)$$

- conjugate base — acid
- or
- base — conjugate acid

Conjugate Acid-Base Pairs

Acid	Base
HF	F⁻
H_2SO_4	HSO_4^-
HSO_4^-	SO_4^{2-}
NH_4^+	NH_3
H_3O^+	H_2O
H_2O	OH^-

H_2O and HSO_4^- are amphoteric, acting as both an acid and a base.

Acid-Base Reactions

- An Arrhenius system is limited to water solutions.

    acid + base → water + salt

- A Brönsted-Lowry system is much more general where a proton is transferred from the acid (HA) to the base (B).

    HA + B ⇌ A$^-$ + BH$^+$

Brönsted-Lowry Acid-Base Reactions

acid + base ⇌ conj. base + conj. acid

HF + NH$_3$ ⇌ F$^-$ + NH$_4^+$

HNO$_2$ + H$_2$O ⇌ NO$_2^-$ + H$_3$O$^+$

H$_2$O + NH$_2^-$ ⇌ OH$^-$ + NH$_3$

Identifying Conjugate Acid-Base Pairs

HNO$_2$ + H$_2$O ⇌ NO$_2^-$ + H$_3$O$^+$

- HNO$_2$ is an acid and NO$_2^-$ is its conjugate base. H$_2$O is a base and H$_3$O$^+$ is its conjugate acid.

    H$_2$O + NH$_2^-$ ⇌ OH$^-$ + NH$_3$

- H$_2$O is an acid and OH$^-$ is its conjugate base. NH$_2^-$ is a base and NH$_3$ is its conjugate acid.

Test Your Skill: Conjugate Pairs

- Write the formula for the conjugate acid of PO_4^{3-}.

- Write the formula for the conjugate base of HClO.

Test Your Skill: Conjugate Pairs

- Identify the conjugate acid-base pairs.

$$HCN(aq) + H_2O(\ell) \rightleftharpoons H_3O^+(aq) + CN^-(aq)$$

$$CH_3NH_2(aq) + H_2O(\ell) \rightleftharpoons CH_3NH_3^+(aq) + OH^-(aq)$$

15.2 Autoionization of Water

- **Objectives**
  - Relate hydrogen ion concentration to hydroxide ion concentration in aqueous solutions.
  - Define pH and use it to express concentrations.
  - Convert between hydrogen ion concentrations, hydroxide ion concentrations, pH, and pOH.

Autoionization of Water

$$H_2O(\ell) + H_2O(\ell) \rightleftharpoons H_3O^+(aq) + OH^-(aq)$$

- H_3O^+ and H_2O are a conjugate acid-base pair.
- H_2O and OH^- are a conjugate acid-base pair.

Autoionization of Water

$$H_2O(\ell) + H_2O(\ell) \rightleftharpoons H_3O^+(aq) + OH^-(aq)$$

- The equilibrium constant for this reaction is denoted as K_w.

  $K_w = [H_3O^+][OH^-]$

- K_w is temperature dependent and is equal to 1.0×10^{-14} at 25 °C.

Example: Autoionization of Water

Calculate the concentration of H_3O^+ in pure water at 25 °C.

Calculating [H$_3$O$^+$] and [OH$^-$]

- Adding an acid or a base to water changes the concentration of both H$_3$O$^+$ and OH$^-$ simultaneously.
- The product of [H$_3$O$^+$] and [OH$^-$] is always equal to K_w.

Example: Calculating [H$_3$O$^+$] and [OH$^-$]

- Calculate the [OH$^-$] in a solution with a [H$_3$O$^+$] of 3.6 × 10^{-3} M.

Test Your Skill: Calculating [H$_3$O$^+$]

- Calculate the hydrogen ion concentration, [H$_3$O$^+$], of a solution in which [OH$^-$] = 0.025 M.
- Is this solution acidic, basic, or neutral?

pH Scale

- **pH** = negative log of the hydronium ion concentration = $-\log_{10}[H_3O^+]$
  - Calculate the pH of a solution that has a 0.0034 M [H_3O^+].

pH units are always given to TWO decimal places.

Example: Calculating [H_3O^+] from pH

- Calculate the hydrogen ion concentration, [H_3O^+], in a solution having a pH of 3.52.

p-notation

- The p-scale can be used for other quantities.

    pOH = $-\log[OH^-]$

    pK_w = $-\log K_w$ = $-\log(1.0 \times 10^{-14})$ = 14.00

pH and pOH Relationship

- $[H_3O^+][OH^-] = K_w = 1.0 \times 10^{-14}$

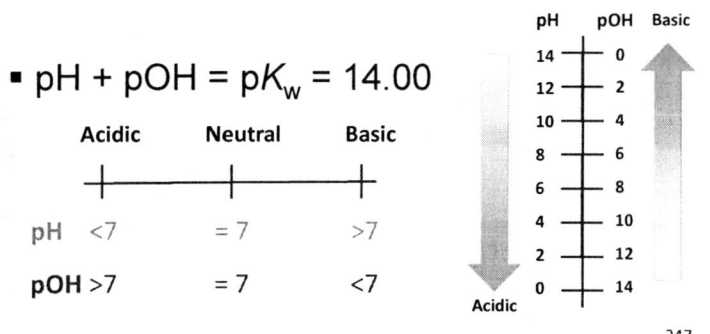

- $pH + pOH = pK_w = 14.00$

	Acidic	Neutral	Basic
pH	<7	=7	>7
pOH	>7	=7	<7

Test Your Skill: Aqueous Solutions

- Complete the first row of the table.

pH	pOH	$[H_3O^+]$	$[OH^-]$	Acidic, Basic, or Neutral?
12.41				
	9.84			
		2.5×10^{-5}		
			2.0×10^{-4}	

- Work the other rows before your next recitation. Your TA will solve the other rows in recitation.

15.3 Strong Acids and Bases

Objectives

- Define strong acids and bases.
- List the species that are strong acids and bases.
- Calculate the concentrations of species, the pH, and the pOH of solutions of strong acids and bases.

Strong Acids

$$HA(aq) + H_2O\;(\ell) \xrightarrow{100\%} H_3O^+(aq) + A^-(aq)$$

- **Strong acids** ionize completely (100%) in solution.

Strong Acids

- There are six common strong acids:

HCl	hydrochloric acid
HBr	hydrobromic acid
HI	hydroiodic acid
HNO_3	nitric acid
$HClO_4$	perchloric acid
H_2SO_4	sulfuric acid*

* $H_2SO_4(aq) + H_2O\;(\ell) \xrightarrow{100\%} H_3O^+(aq) + HSO_4^-(aq)$

Example: The pH of Strong Acids

- What is the pH of a 0.050 *M* HCl solution?

Strong Bases

- **Strong bases** quantitatively produce hydroxide ions in water.
- Common strong bases include the group IA, Sr, and Ba hydroxides and oxides.

$$NaOH(s) \rightarrow Na^+(aq) + OH^-(aq)$$
$$Li_2O(s) + H_2O(\ell) \rightarrow 2Li^+(aq) + 2OH^-(aq)$$

Example: pH of a Strong Base

- Calculate the pH of a 0.035 M barium hydroxide solution.

Test Your Skill: Calculate Concentration

- Calculate the concentration of a HCl solution having a pH of 3.75.
- Calculate the concentration of a Sr(OH)$_2$ solution having a pH of 11.60.

15.4 Qualitative Aspects of Weak Acids and Weak Bases

- **Objectives**
  - Define weak acids and bases
  - Relate K_a to the competition of different bases for protons

Weak Acids and Bases

- **Weak acids** and **weak bases** do not ionize completely in water.
- A weak acid or base exists in equilibrium with its conjugate partner.

$$HCN(aq) + H_2O(\ell) \rightleftharpoons H_3O^+(aq) + CN^-(aq)$$
$$CH_3NH_2(aq) + H_2O(\ell) \rightleftharpoons CH_3NH_3^+(aq) + OH^-(aq)$$

Weak Acids and Bases (General Forms)

$$HA(aq) + H_2O(\ell) \rightleftharpoons H_3O^+(aq) + A^-(aq)$$

$$K_a = \frac{[H_3O^+][A^-]}{[HA]}$$

$$B(aq) + H_2O(\ell) \rightleftharpoons BH^+(aq) + OH^-(aq)$$

$$K_b = \frac{[BH^+][OH^-]}{[B]}$$

Competition for Protons: Weak Acids

- Both H_2O and F^- are bases and compete for protons (H^+).

  $HF(aq) + H_2O(\ell) \rightleftharpoons H_3O^+(aq) + F^-(aq)$

- Because HF is a weak acid, the reaction reaches equilibrium. F^- is a weak base and will accept protons easily.

Competition for Protons: Strong Acids

$HCl(aq) + H_2O(\ell) \xrightarrow{100\%} H_3O^+(aq) + Cl^-(aq)$

- Because HCl is a strong acid, the reaction will proceed in only one direction and Cl^- does not behave as a base.
- Cl^- is a spectator ion in acid-base reactions.

15.5 Weak Acids

- **Objectives**
  - Write the chemical equation for the ionization of a weak acid.
  - Define analytical concentration.
  - Relate the fraction of a weak acid ionized to the acid ionization constant.

15.5 Weak Acids

- **Objectives continued**
  - Calculate acid ionization constants from experimental data.
  - Calculate the concentrations of the species present in a weak acid solution.
  - Calculate percent ionization from K_a and concentration.

Calculating K_a for a Weak Acid

- To calculate the equilibrium constant K_a for a weak acid we will use the same method as we did for calculating K_{eq}.

$$HF(aq) + H_2O(\ell) \rightleftharpoons H_3O^+(aq) + F^-(aq)$$

$$K_a = \frac{[H_3O^+][F^-]}{[HF]}$$

Example: Calculating K_a

- The pH of a 0.100 M solution of the weak acid HOCl is 4.20. Calculate K_a for the acid.

Test Your Skill: Calculate K_a

- The pH of a 0.50 M solution of HOCl is 3.85. Calculate K_a for HOCl.

Example: Calculating Equilibrium Concentrations

- Calculate the pH of a 0.50 M acetic acid (CH$_3$COOH) solution. $K_a = 1.8 \times 10^{-5}$

Test Your Skill: Weak Acid pH

- Calculate the pH of a 0.025 M solution of HCN, $K_a = 7.2 \times 10^{-10}$.

Analytical Concentration of an Acid

- **Analytical concentration**: the total concentration of all forms of an acid; both the protonated and unprotonated forms.
- HA(aq) + H$_2$O(ℓ) $\rightleftharpoons$ H$_3$O$^+$(aq) + A$^-$(aq)
- The analytical concentration of HA is equal to the sum of the equilibrium concentrations of HA and A$^-$.
- The analytical concentration is the initial concentration in the iCe table.

Ionic Concentration as a Function of Analytical Concentration

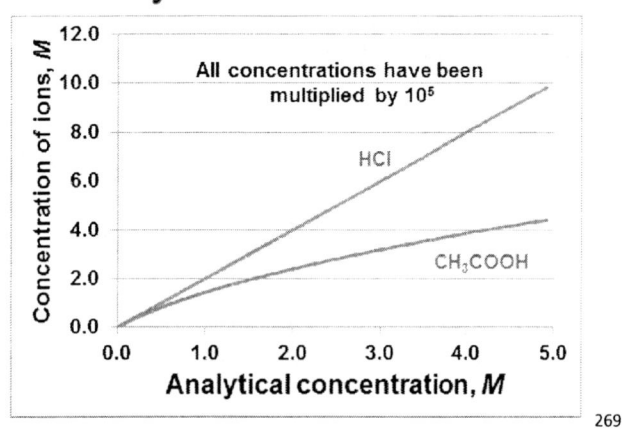

Fraction Ionized

HA(aq) + H$_2$O(ℓ) $\rightleftharpoons$ H$_3$O$^+$(aq) + A$^-$(aq)

$$\text{fraction ionized} = \frac{[A^-]}{[HA]+[A^-]}$$

Since [H$_3$O$^+$] = [A$^-$]

$$\text{fraction ionized} = \frac{[H_3O^+]}{[HA]+[A^-]}$$

$$\text{fraction ionized} = \frac{[H_3O^+]}{\text{analytical conc. of HA}}$$

Fraction ionized is expressed as a percentage.

Example: Fraction Ionized

- Calculate the fraction ionized for 0.500 M HOCl, $K_a = 4.0 \times 10^{-8}$.

Test Your Skill: Fraction Ionized

- Calculate fraction of HOCl ionized in 0.0500 M HOCl, $K_a = 4.0 \times 10^{-8}$.

15.6 Solutions of Weak Bases and Salts

- **Objectives**
  - Write the chemical reaction that occurs when a weak base dissolves in water.
  - Calculate the concentrations of species present in a solution of a weak base.
  - Relate K_b for a base to K_a of its conjugate acid.
  - Calculate the pH of a salt solution.

Solutions of Weak Bases

- A weak base reacts with water to form hydroxide ions [OH⁻].

$$B(aq) + H_2O(\ell) \rightleftharpoons BH^+(aq) + OH^-(aq)$$

$$K_b = \frac{[BH^+][OH^-]}{[B]}$$

Calculating the pH of a Weak Base

- Calculate the pH of a 0.150 M methylamine (CH_3NH_2) solution. $K_b = 4.4 \times 10^{-4}$

Test Your Skill: Weak Base pH

- Write the equation for the reaction of the weak base hydroxylamine (NH_2OH, $K_b = 1.1 \times 10^{-8}$) in water and calculate the pH of a 0.35 M solution.
- Hint: A weak base accepts a proton.

Relating K_a to K_b

$$HA(aq) + H_2O(\ell) \rightleftharpoons H_3O^+(aq) + A^-(aq)$$

$$K_a = \frac{[H_3O^+][A^-]}{[HA]}$$

$$A^-(aq) + H_2O(\ell) \rightleftharpoons HA(aq) + OH^-(aq)$$

$$K_b = \frac{[HA][OH^-]}{[A^-]}$$

Relating K_a to K_b

$$K_a \times K_b = \frac{[H_3O^+][A^-]}{[HA]} \times \frac{[HA][OH^-]}{[A^-]}$$

$$K_a \times K_b = [H_3O^+][OH^-]$$

$$K_a \times K_b = K_w$$

Remember: The K_a and K_b refer to an acid-base conjugate pair.

Strengths of Weak Acid-Base Conjugate Pairs

- The conjugate base of a weak acid is a weak base:

Acid	Conjugate base
CH_3COOH	CH_3COO^-
$K_a = 1.8 \times 10^{-5}$	$K_b = 5.6 \times 10^{-10}$

$$K_a \times K_b = K_w$$

$$(1.8 \times 10^{-5}) \times (5.6 \times 10^{-10}) = 1.0 \times 10^{-14}$$

Example: Calculating K_b from K_a

- K_a for HNO_2 is 4.6×10^{-4}. Identify the conjugate base of HNO_2, write its reaction in water, and calculate K_b.

Test Your Skill: Calculating K_a from K_b

- K_b for pyridine (C_5H_5N) is 1.8×10^{-9}. Identify the conjugate acid of pyridine, write its reaction in water, and calculate K_a.

Acid-Base Strengths

- The strengths of weak acids and bases are related to the value of their equilibrium constants.
  - The larger the K_a, the stronger the acid.
  - The larger the K_b, the stronger the base.
- There are no values of K_a or K_b for aqueous strong acids and bases, like HCl and NaOH.
  - The conjugate partners Cl⁻ and Na⁺ have no acid-base properties.

Weak Acid-Base Strengths

- The stronger the acid, the weaker its conjugate base.
  - HF is a stronger acid than HCN.
  - F⁻ is a weaker base than CN⁻.

HF $K_a = 6.3 \times 10^{-4}$ F⁻ $K_b = 1.6 \times 10^{-11}$
HCN $K_a = 6.2 \times 10^{-10}$ CN⁻ $K_b = 1.6 \times 10^{-5}$

Estimating the pH of Salt Solutions

$$MX(s) \rightarrow M^+(aq) + X^-(aq)$$

- **Strategy:** Because all soluble salts dissociate into ions, we need to look at the acid-base properties of the cation and the anion.
  - Anion or cation is the conjugate partner of a strong acid or base: it is a spectator ion and has no acid-base properties.
  - Anion or cation is the conjugate partner of a weak acid or base: it has acid-base properties.

pH of Salt Solutions

$$MX(s) \rightleftharpoons M^+(aq) + X^-(aq)$$

Example	Cation	Anion	pH of Solution
NH₄Cl	NH₄⁺ (conj. acid of NH₃)	Cl⁻ (spectator)	**Acidic**
NaF	Na⁺ (spectator)	F⁻ (conj. base of HF)	**Basic**
NaCl	Na⁺ (spectator)	Cl⁻ (spectator)	Neutral
NH₄F	NH₄⁺ (conj. acid of NH₃)	F⁻ (conj. base of HF)	Compare K_a and K_b

Test Your Skill: Estimating pH

- Estimate the pH of aqueous solutions of the following.
  - Sulfuric acid
  - Barium hydroxide
  - Lactic acid
  - Ammonia
  - Sodium acetate

  > Choose from:
  > Strong acid = 1
  > Weak acid = 4
  > Neutral = 7
  > Weak base = 10
  > Strong base = 13

Example: pH of Salt Solutions

- Calculate the pH of a 0.59 M solution of sodium acetate (CH_3COONa). K_a for $CH_3COOH = 1.8 \times 10^{-5}$.

15.7 Mixtures of Acids

- **Objectives**
  - Calculate the pH of a solution that contains both strong and weak acids.
  - Determine the pH of a solution that contains a mixture of weak acids (or bases) with different ionization constants.

Mixtures of Acids

- When multiple strong acids are present, the effects are additive.
- When a strong acid and a weak acid are present, ignore the weak acid.
- When multiple weak acids are present, consider only the weak acid with largest value of K_a if the acid ionization constants differ by a factor of 100 or more.

Example: Mixtures of Acids

- Calculate the pH of a solution that is:
    0.50 M HF (K_a for HF is 6.3×10^{-4})
    0.10 M HCl
    0.20 M HBr

15.8 Influence of Molecular Structure on Acid Strength

- **Objectives**
  - Relate the acid ionization constants of a series of related binary acids to their structure and bonding.
  - Define oxyacid and list several common oxyacids.
  - Explain how fundamental properties such as size and electronegativity affect the strengths of acids.

Binary Hydrides

- **Binary hydrides** contain hydrogen and one other element.
- Acidity depends on the HA bond strength.

Binary Hydrides

- The acidity of a compound increases as the H-A bond dissociation energy decreases. The weaker the H-A bond the more acidic the acid.

Acid	Bond Energy, kJ/ mol
HI	298.3
HBr	366.1
HCl	431.9
HF	568.2

Acid strength ↑   Bond energy ↓

Binary Hydrides

- As the electronegativity of A in the binary hydride HA increases the acidity increases

Acid	Electronegativity
HF	4.0
H_2O	3.5
NH_3	3.0
CH_4	2.5

Acid strength ↑   Electronegativity ↑

Oxyacids

- **Oxyacids** contain H, O and a third element, X.
  - Examples: HNO_2, HNO_3, and H_2CrO_4
- X is usually a non-metal or a transition metal in a high oxidation state.

Oxyacids

- The two factors that increase the acidity of oxyacids are:
  - Greater electronegativity of the central atom

    $HBrO_4$ is more acidic than HIO_4.
  - Greater oxidation number of the central atom.

    $HClO_4 > HClO_3 > HClO_2 > HClO$

15.9 Lewis Acids and Bases

- **Objectives**
  - Define Lewis acids and bases.
  - Identify Lewis acids and bases and their reaction products.

Lewis Acids and Bases

- **Lewis acid**: electron pair acceptor.
  - Boron trichloride, BCl_3, acts as Lewis acid because it has a vacant orbital to accept electrons.
- **Lewis base**: electron pair donor.
  - Ammonia, NH_3, acts as a Lewis base because it has a lone pair of electrons.

Lewis Acid-Base Reactions

- **Lewis acid-base reaction**: formation of a coordinate-covalent bond.
- **Coordinate-covalent bond**: covalent bond in which both electrons come from one atom.

$$AlCl_3 + Cl^- \rightarrow AlCl_4^-$$

Chapter 16

Reactions Between Acids and Bases

16.1 Titrations of Strong Acids and Bases

- **Objectives**
  - Define titration.
  - Calculate the volume of titrant needed to reach the equivalence point.
  - Identify regions of the titration curve in which the analyte or titrant is in excess.
  - Express amounts of analyte and titrant in units of millimoles.

Titration of Strong Acids and Bases

- **Titration**: a method used to determine the concentration of a substance called the analyte by adding another substance called the titrant, which reacts in a known manner with the analyte.

Titration of Strong Acids and Bases

- A known volume of an acid with unknown concentration is added to a flask.
- A standardized base is added using a buret.
- The endpoint is indicated by a color change.
- The volume of the base added to reach the endpoint is recorded.

Titration of Strong Acids and Bases

- **Titration Curve**: a graph of pH of a solution versus volume of titrant added.
- Reactions of strong acids or bases go to completion.

  HCl(aq) + NaOH(aq) → H$_2$O(ℓ) + NaCl(aq)

- For a titration of a strong acid, the pH starts low and stays low as long as strong acid is present.

Titration of Strong Acids and Bases

- The pH rises sharply to 7 at the **equivalence point**, where the acid and base are present in stoichiometrically equivalent amounts.
- After excess strong base has been added, the pH levels off at a high value.

Titration Curve of a Strong Acid with a Strong Base

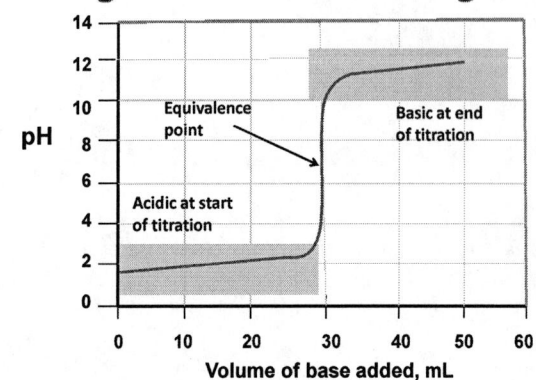

Example: Equivalence Point Calculation

- Calculate the equivalence point volume in the titration of 20.00 mL of 0.1252 M HCl with 0.1008 M NaOH.

Test Your Skill: Titration

- Calculate the equivalence point volume in the titration of 40.00 mL of 0.2387 M HNO_3 with 0.3255 M $Ba(OH)_2$.

Millimoles

- **Millimole, (mmol):** one thousandth $\left(\frac{1}{1000}\right)$ of a mole.
- Units of Molarity:
- $1.0\ M = 1.0\ \frac{mol}{L} = 1.0\ \frac{mmol}{mL}$
- $mmol = mL\left(\frac{mmol}{mL}\right)$

Example: Millimoles

- Calculate the number of mmol of hydroxide ion in 15.00 mL of 0.100M strontium hydroxide.

16.2 Titration Curves of Strong Acids and Bases

- **Objectives**
  - Calculate the concentrations of all species present during the titration of a strong acid with a strong base.
  - Graph the titration curve.
  - Correlate the shape of the titration curve to the titration stoichiometry.
  - Estimate the pH of mixtures of strong acids and bases.

Calculating a Titration Curve

- Write the chemical reaction for the titration.
- Determine the mmol of acid and base from the product of the volume and the concentrations.
- Place information into a sRfc table.
- Determine from table the **concentration** of excess H_3O^+ or OH^-.
- Calculate pH.

sRfc Table

	$H_3O^+(aq)$	$+ OH^-(aq)$	$\rightarrow 2H_2O\ (\ell)$
s - starting amount, mmol			
R - reacting amount, mmol			
f - final amount, mmol			
c - final concentration, M			

Example: Calculating a Titration Curve

- Calculate the pH in the titration of 20.0 mL of 0.125 M HCl with 0.250 M NaOH after 0, 2.00, 10.0 and 20.0 mL of base are added.
- There is space in the lecture notes to work the problem.

Calculating a Titration Curve

0.00 mL of base added to 20.0 mL 0.125 M HCl:

Calculating a Titration Curve

2.00 mL 0.250 M NaOH added to 20.0 mL 0.125 M HCl:

	H_3O^+(aq) +	OH^-(aq) →	$2H_2O$ (ℓ)
s, mmol			
R, mmol			
f, mmol			
c, M			

Calculating a Titration Curve

10.0 mL 0.250 M NaOH added to 20.0 mL 0.125 M HCl:

	H_3O^+(aq) +	OH^-(aq) →	$2H_2O$ (ℓ)
s, mmol			
R, mmol			
f, mmol			
c, M			

Calculating a Titration Curve

20.0 mL 0.250 M NaOH added to 20.0 mL 0.125 M HCl:

	H_3O^+(aq) +	OH^-(aq) →	$2H_2O$ (ℓ)
s, mmol			
R, mmol			
f, mmol			
c, M			

Test Your Skill: Titration

- Calculate two more pH data points and graph the full titration curve.
  - 9.90 mL 0.250 *M* NaOH added to 20.0 mL 0.125 *M* HCl
  - 10.1 mL 0.250 *M* NaOH added to 20.0 mL 0.125 *M* HCl

Test Your Skill: Titration

9.90 mL 0.250 *M* NaOH added to 20.0 mL 0.125 *M* HCl:

	$H_3O^+(aq)$ +	$OH^-(aq)$	$\rightarrow 2H_2O\ (\ell)$
s, mmol			
R, mmol			
f, mmol			
c, *M*			

Test Your Skill: Titration

10.1 mL 0.250 *M* NaOH added to 20.0 mL 0.125 *M* HCl:

	$H_3O^+(aq)$ +	$OH^-(aq)$	$\rightarrow 2H_2O\ (\ell)$
s, mmol			
R, mmol			
f, mmol			
c, *M*			

Test Your Skill: Titration

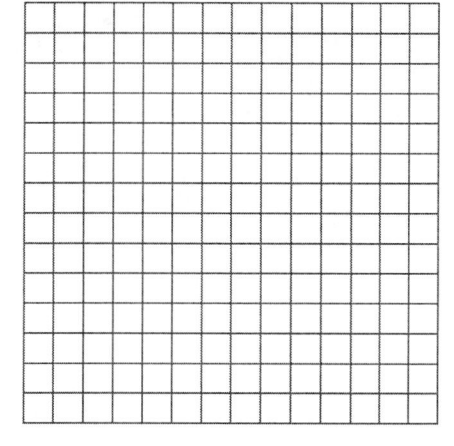

pH

Volume NaOH, mL

Titration Curve of 50.00 mL of 0.500 M NaOH with 1.00 M HCl

Volume of acid added	pH
0.00	13.70
10.00	13.40
24.00	12.15
25.00	7.00
26.00	1.88
40.00	0.78

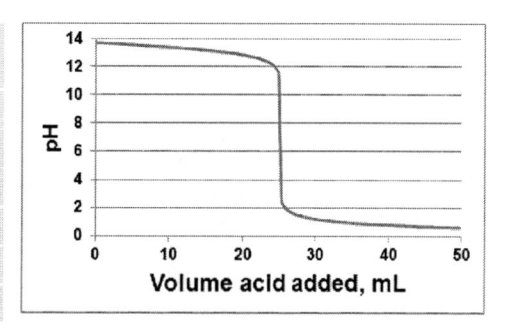

Stoichiometry and Titrations
- The equivalence point volume is related to the stoichiometry.

$$H_2SO_4 + 2NaOH \rightarrow Na_2SO_4 + 2H_2O$$
$$HCl + NaOH \rightarrow NaCl + H_2O$$

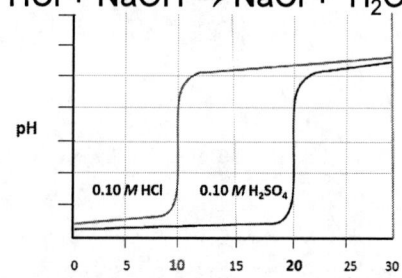

pH

0.10 M HCl   0.10 M H$_2$SO$_4$

Volume of 0.10 M NaOH added to 10.0 mL of acid

Example: 1:2 Stoichiometry

- Calculate the pH when 10.0 mL of 0.35M sodium hydroxide solution are added to 10.0 mL of 0.10M sulfuric acid solution.

Example: Estimate the pH of Mixtures of Strong Acids and Bases

- Estimate the pH of a solution that contains 10.0 mL 0.20M HCl, 15.0 mL 0.10M HNO_3, and 10.0 mL 0.30M NaOH.

16.3 Buffers

- **Objectives**
  - Describe the function and composition of a pH buffer.
  - Calculate the pH of a buffer solution from the concentrations of the weak acid and its conjugate base.

16.3 Buffers

- **Objectives continued**
  - Calculate the pH of a solution from the amounts of acids and bases.
  - Determine the change in the pH when strong acid or base is added to a buffer.

Buffers

- **Buffer**: a solution that resists changes in pH.
- A **buffer** is a mixture of a weak acid or a weak base and its conjugate partner.

$$HA \,/\, A^- \quad \text{or} \quad B \,/\, BH^+$$
$$HF \,/\, F^- \quad \text{or} \quad NH_3 \,/\, NH_4^+$$

Buffers

Any added OH^- reacts with the weak acid

$$OH^-_{(aq)} + HF_{(aq)} \rightarrow H_2O_{(\ell)} + F^-_{(aq)}$$

$HF_{(aq)} \quad F^-_{(aq)}$
$F^-_{(aq)} \quad HF_{(aq)}$

Any added H_3O^+ reacts with the weak base

$$H_3O^+_{(aq)} + F^-_{(aq)} \rightarrow H_2O_{(\ell)} + HF_{(aq)}$$

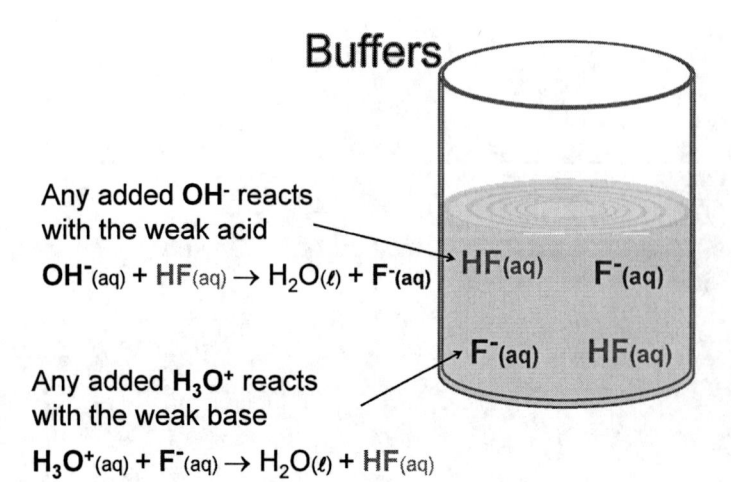

pH of Buffer Solutions

For the chemical equilibrium:
$$HF + H_2O \rightleftharpoons H_3O^+ + F^-$$

$$K_a = \frac{[H_3O^+][F^-]}{[HF]}$$

$$[H_3O^+] = \frac{K_a[HF]}{[F^-]}$$

Taking the –log (p – function) of both sides gives:

$$pH = pK_a + \log\frac{[F^-]}{[HF]}$$

Henderson-Hasselbalch Equation

- $pH = pK_a + \log\dfrac{C_b}{C_a}$

    C_b = concentration of weak base, A⁻

    C_a = concentration of weak acid, HA

Example: pH of a Buffer Solution

- Calculate the pH of a solution of 0.50 M HCN and 0.20 M NaCN, $K_a = 4.9 \times 10^{-10}$.

Example: pH of a Buffer Solution

- Calculate the pH of a solution of 0.40 M NH_3 and 0.10 M NH_4Cl, $K_b = 1.8 \times 10^{-5}$.

Test Your Skill: Buffers

- Calculate the pH of a buffer that is 0.25 M HCN and 0.15 M NaCN, $K_a = 4.9 \times 10^{-10}$.

pH of a Buffer System

- The Henderson-Hasselbalch equation can be rewritten as:

$$pH = pK_a + \log \frac{n_b}{n_a}$$

- Where n_b and n_a are the number of moles or millimoles of the base and acid, respectively.

Example: Composition of a Buffer System

- Calculate the mass of sodium acetate that must be added to 250 mL of 0.16 M acetic acid in order to prepare a pH 4.68 buffer. $K_a = 1.8 \times 10^{-5}$. Molar mass of CH_3COONa is 82.0 g/mol.

Test Your Skill: Composition of a Buffer System

- How many moles of NaCN should be added to 100 mL of 0.25 M HCN to prepare a buffer with a pH = 9.40? $K_a = 4.9 \times 10^{-10}$

Example: Response of a Buffer to Added Acid or Base

- Calculate the initial and final pH when 10.0 mL of 0.100 M HCl is added to

(a) 100 mL of water

(b) 100 mL of a buffer that is 1.50 M CH_3COOH and 1.20 M CH_3COONa. $K_a = 1.8 \times 10^{-5}$.

Test Your Skill: Response of a Buffer to Added Acid or Base

- Calculate the final pH when 10.0 mL of 0.100 M NaOH is added to 100 mL of a buffer which is 1.50 M CH_3COOH and 1.20 M CH_3COONa. $K_a = 1.8 \times 10^{-5}$.

16.4 Titrations of Weak Acids and Bases: Qualitative Aspects

- **Objectives**
  - Separate the titration curve for a weak acid into regions in which a single equilibrium dominates
  - Estimate the pH of mixtures of a weak acid and strong base or weak base and strong acid.

Titration: Weak Acid + Strong Base

- Strategy:
  - pH before any base is added.
  - pH after base is added but before equivalence point.
  - pH at equivalence point.
  - pH after equivalence point.

Titration: Weak Acid + Strong Base

- pH before any base is added.
  - The solution is a weak acid.
  - Use K_a and an iCe table to solve for pH.

Titration: Weak Acid + Strong Base

- pH after base is added but before equivalence point.
  - The strong base is the limiting reactant.
  - The resulting solution is a mixture of the weak acid and its conjugate base – a buffer.
  - The pH can be determined using an sRf table and the Henderson-Hasselbalch equation.

Titration: Weak Acid + Strong Base

- pH at equivalence point.
  - All the strong base is completely consumed.
  - All of the weak acid is consumed and the conjugate base forms.
  - The sRfc table show the resulting solution is a weak base.
  - Use K_b and an iCe table to solve for pH.

Titration: Weak Acid + Strong Base

- pH after equivalence point.
  - The weak acid is the limiting reactant.
  - The sRfc table shows the resulting solution is a strong base.
  - The pH can be determined from the pOH.

Titration: Weak Acid + Strong Base

pH vs. Volume of NaOH added, mL

Labels: Buffer region, Weak acid, Midpoint of titration, Equivalence point, Strong base region

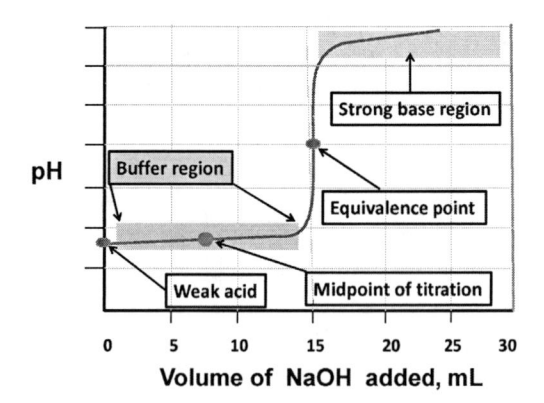

Titration Curves for Same Amounts of Different Strength Acids

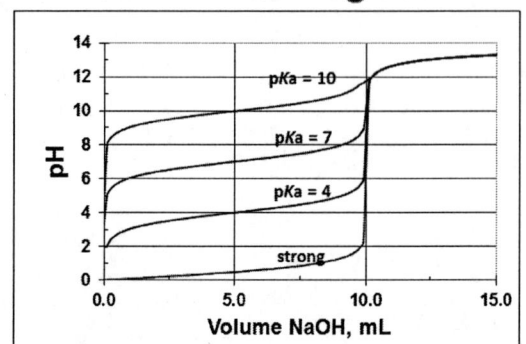

Curves shown for pKa = 10, pKa = 7, pKa = 4, strong

Note: It takes the same amount of base to neutralize 1 mmol of strong acid as 1 mmol of weak acid.

16.5 Titrations of Weak Acids and Bases: Quantitative Aspects

- **Objectives**
  - Calculate the pH in the titration of a weak acid with strong base.
  - Calculate the pH in the titration of a weak base with strong acid.

Example: Titration Curve for a Weak Acid

- Calculate the pH in the titration of 20.00 mL of 0.500 M formic acid (K_a = 1.8 × 10^{-4}) with 0.500 M NaOH after 0, 10.00, 20.00, and 30.00 mL of base have been added. Graph the titration curve.

Example: Titration Curve for a Weak Acid

0 mL 0.500 M NaOH + 20.0 mL 0.500 M HCOOH

Example:
Titration Curve for a Weak Acid

10.00 mL 0.500 M NaOH + 20.0 mL 0.500 M HCOOH

	HCOOH +	OH⁻ →	HCOO⁻ +	H₂O
s, mmol				
R, mmol				
f, mmol				
c, M				

Example:
Titration Curve for a Weak Acid

20.00 mL 0.500 M NaOH + 20.0 mL 0.500 M HCOOH

	HCOOH +	OH⁻ →	HCOO⁻ +	H₂O
s, mmol				
R, mmol				
f, mmol				
c, M				

Example:
Titration Curve for a Weak Acid

30.00 mL 0.500 M NaOH + 20.0 mL 0.500 M HCOOH

	HCOOH +	OH⁻ →	HCOO⁻ +	H₂O
s, mmol				
R, mmol				
f, mmol				
c, M				

Titration Curve for a Weak Acid

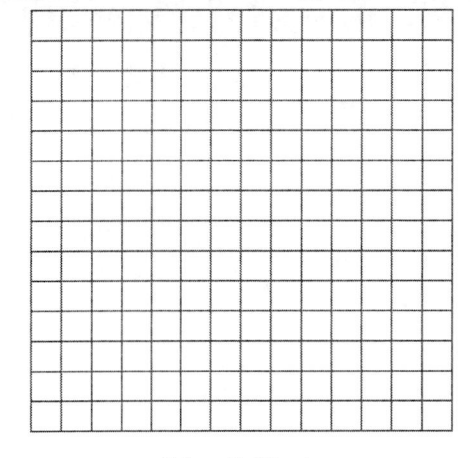

pH

Volume NaOH, mL

Test Your Skill: Weak Acid Titration

- Calculate the pH in the titration of 12.00 mL of 0.100 M HOCl with 0.200 M NaOH after 0, 3.00, 6.00 and 9.00 mL of base have been added. Graph the titration curve. $K_a = 4.0 \times 10^{-8}$

12.00 mL of 0.100 M HOCl with 0 mL 0.200 M NaOH

12.00 mL of 0.100 M HOCl with
3.00 mL 0.200 M NaOH

12.00 mL of 0.100 M HOCl with
6.00 mL 0.200 M NaOH

12.00 mL of 0.100 M HOCl with
9.00 mL 0.200 M NaOH

Graph the Titration Curve

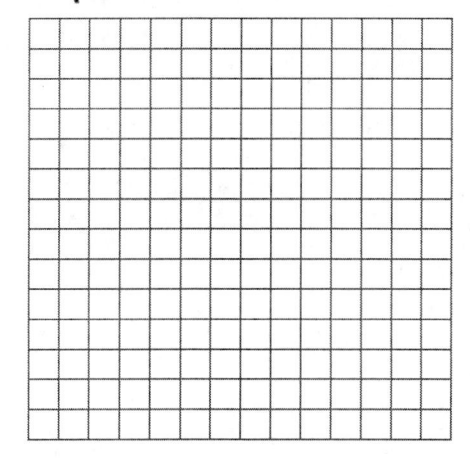

Titration of 20.00 mL of 0.500 M Methylamine (K_b = 3.7 × 10^{-4}) with 0.500 M HCl

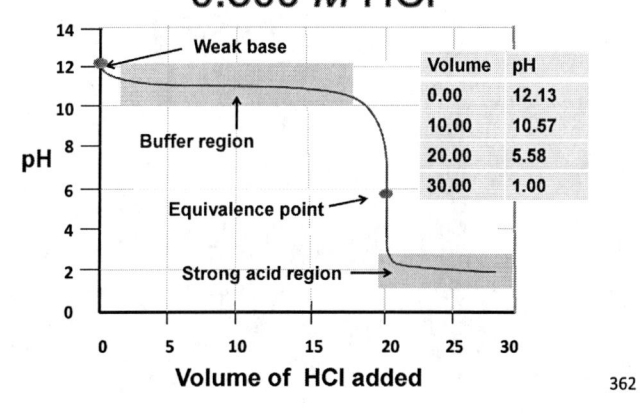

Volume	pH
0.00	12.13
10.00	10.57
20.00	5.58
30.00	1.00

16.6 Indicators

- **Objectives**
  - Describe indicators by their acid-base chemistry.
  - Choose an indicator that is appropriate for a particular titration.

pH Indicators

- **Indicator**: a substance that changes color at the endpoint of a titration.
- pH indicators are weak acids or bases whose conjugate partners are a different color in solution.

pH Indicators

$$HIn + H_2O \rightleftharpoons H_3O^+ + In^-$$

$$K_{In} = \frac{[H_3O^+][In^-]}{[HIn]}$$

$$pK_{In} = -\log(K_{In})$$

pH Indicators

- When pH is lower than pK_{In}, the indicator will be in the acid form.
- When pH is greater than pK_{In}, the indicator will be in the base form.
- An indicator should be chosen so that the pH of the color change is at or just beyond the titration equivalence point pH.

Examples of pH Indicators

Name	Acid Color	Base Color	pH Range	pK_{In}
Thymol blue*	Red	Yellow	1.2–2.8	1.6
Methyl orange	Red	Yellow	3.1–4.4	3.5
Methyl red	Red	Yellow	4.2–6.3	5.0
Bromthymol blue	Yellow	Blue	6.2–7.6	7.3
Phenolphthalein	Colorless	Pink	8.3–10.0	8.7
Thymol blue*	Yellow	Blue	8.0–9.6	9.2

*Thymol blue is polyprotic and has three color forms.

Example: Choose an Indicator

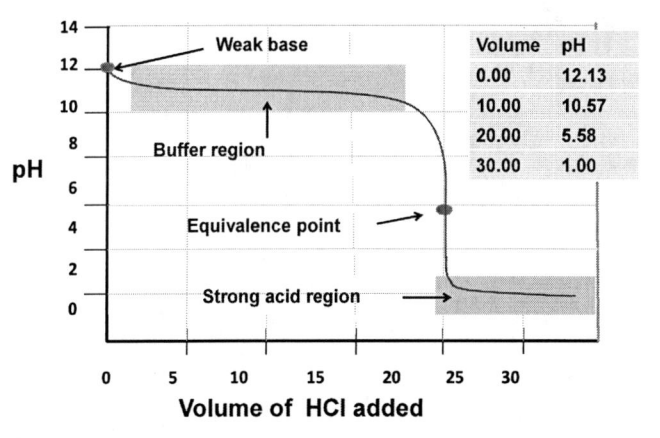

Volume	pH
0.00	12.13
10.00	10.57
20.00	5.58
30.00	1.00

Weak base
Buffer region
Equivalence point
Strong acid region

Volume of HCl added

16.7 Polyprotic Acid Solutions

- **Objective**
  - Write chemical equations and expressions for the equilibrium constants for the dissociation of polyprotic acids

Polyprotic Acids

- **Polyprotic acids** provide more than one proton when they ionize.
- Polyprotic acids ionize in a stepwise manner.

$H_2A + H_2O \rightleftharpoons H_3O^+ + HA^-$ Step 1

$HA^- + H_2O \rightleftharpoons H_3O^+ + A^{2-}$ Step 2

Polyprotic Acids

- There is a separate acid ionization constant for each step.

$H_2A + H_2O \rightleftharpoons H_3O^+ + HA^-$ Step 1

$$K_{a1} = \frac{[H_3O^+][HA^-]}{[H_2A]}$$

$HA^- + H_2O \rightleftharpoons H_3O^+ + A^{2-}$ Step 2

$$K_{a2} = \frac{[H_3O^+][A^{2-}]}{[HA^-]}$$

Polyprotic Acids

- HA^- is the conjugate base of H_2A, so it is a weaker acid than H_2A.
  - K_{a1} is always larger than K_{a2}
- For triprotic acids like H_3PO_4.
  - $K_{a1} > K_{a2} > K_{a3}$

Amphoteric Species

- An **amphoteric species** is one that can have both acidic and basic properties.
- Conjugate bases of weak polyprotic acids are amphoteric. The conjugate base of carbonic acid (H_2CO_3), hydrogen carbonate ion (HCO_3^-), is **amphoteric**.
  - $H_2CO_3 + H_2O \rightleftharpoons \mathbf{HCO_3^-} + H_3O^+$
  - $\mathbf{HCO_3^-} + H_2O \rightleftharpoons CO_3^{2-} + H_3O^+$

Amphoteric Species

1. $H_2CO_3 + H_2O \rightleftharpoons HCO_3^- + H_3O^+$
   $K_{a1} = 4.5 \times 10^{-7}$
2. $HCO_3^- + H_2O \rightleftharpoons CO_3^{2-} + H_3O^+$
   $K_{a2} = 4.7 \times 10^{-11}$

In order to estimate the pH of an amphoteric species, we compare K_a to K_b.
Equation 2 shows HCO_3^- as a weak acid.
$K_a = 4.7 \times 10^{-11}$.
Equation 1 shows HCO_3^- as a weak base.
$K_b = K_w/K_{a1} = 1.0 \times 10^{-14}/4.5 \times 10^{-7} = 2.2 \times 10^{-8}$

Amphoteric Species

- As an acid: $K_a = 4.7 \times 10^{-11}$
- As a base: $K_b = 2.2 \times 10^{-8}$
- Since $K_b \gg K_a$ a hydrogen carbonate solution is weakly basic.

Test Your Skill: Amphoteric Species

- Is a solution of sodium hydrogen malonate acidic or basic in water? Malonic acid is diprotic and can be represented with the formula H_2A. $K_{a1} = 1.4 \times 10^{-3}$ and $K_{a2} = 2.0 \times 10^{-6}$.

16.8 Factors That Influence Solubility

- **Objective**
  - Determine how pH influences the solubility of precipitates

Factors That Influence Solubility

- The solubilities of salts in which the anion is a weak base – for example, F^-, CH_3COO^-, CO_3^{2-} – are affected by pH.

Salts in Which the Anion is a Weak Base

- The solubility of salts in which the anion is a weak base can be predicted from LeChatelier's principle.

$$Cd(CN)_2(s) \rightleftharpoons Cd^{2+}(aq) + 2CN^-(aq)$$

- Adding acid reduces [CN^-] in solution:

$$H_3O^+(aq) + CN^-(aq) \rightarrow HCN(aq) + H_2O(\ell)$$

- Removing CN^- causes the solubility equilibrium to proceed forward.
- The solubility increases when acid is added.

Chapter 17

Chemical Thermodynamics

Chemical Thermodynamics

- **Chemical thermodynamics**: the study of the energy changes during the course of a chemical reaction.

Review

- The **system** is the part of the universe under examination. The chemical reaction is the system.
- The **surroundings** are the rest of the universe. The beaker and the lab bench are part of the surroundings.
- The system plus the surroundings equals the **universe**.
- The **change in enthalpy (ΔH)** is the heat absorbed (ΔH positive) or released (ΔH negative) by the system at constant temperature and pressure.

17.1 Work and Heat

Objectives

- Define energy, heat, and work.
- Relate the signs of heat and work to the processes that occur.
- Calculate work from pressure-volume relationships.

Energy, Heat and Work

- The word **energy** derives from the Ancient Greek *energeia*, meaning activity or operation.
- Under most laboratory conditions, a system exchanges energy with the surroundings as heat and/or work.
- **Heat, q,** is random energy, and is associated with the temperature of the system.
- **Work, w,** is directed energy, and is associated with the moving part of the system.

Heat

- Calorimetry measures heat transfer to the surroundings.

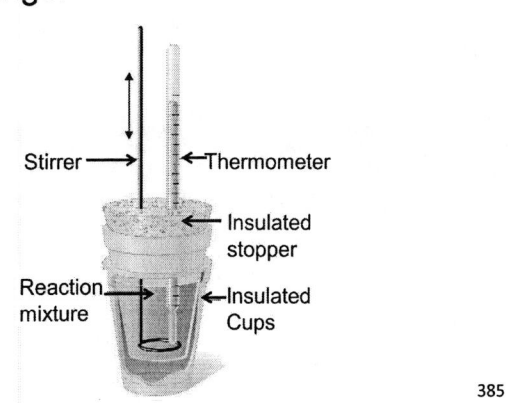

Sign of Heat

- The sign of q is centered around the system.

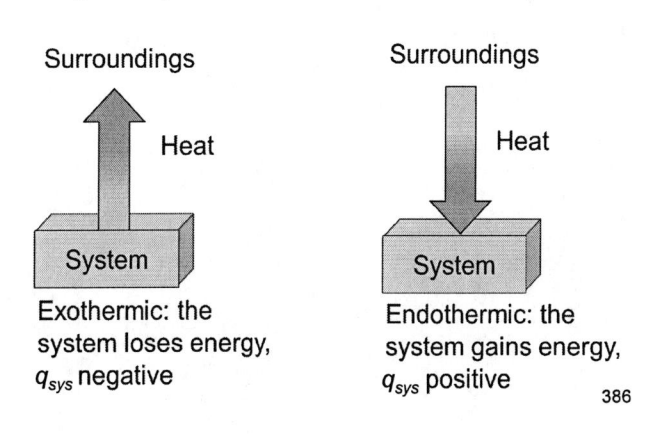

Exothermic: the system loses energy, q_{sys} negative

Endothermic: the system gains energy, q_{sys} positive

Work

Work is an application of a force through a distance.
work = force × distance

When the system does work, like lifting a weight against gravity, the sign for work is negative.

Work Done by Gases

A pressure difference causing a volume change is also work.
$w = -P\Delta V$
w can be expressed in L·atm or J.
1 L·atm = 101.3 J

Example: Work (qualitative)

- Describe what happens to the pressure and volume of the systems below.
- Predict the signs of w.

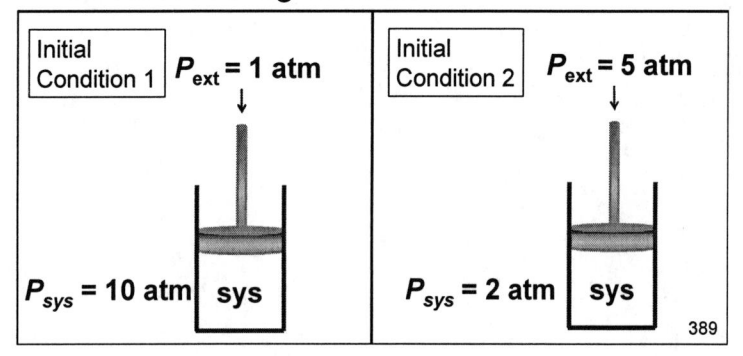

Initial Condition 1: P_{ext} = 1 atm, P_{sys} = 10 atm

Initial Condition 2: P_{ext} = 5 atm, P_{sys} = 2 atm

Example: Pressure-Volume Work

- Express the work (in joules) when 20.0 L of an ideal gas at a pressure of 12.0 atm expands against a constant pressure of 1.5 atm. Assume constant temperature.
    1 L·atm = 101.3 J

Review of Sign Conventions for q and w

- The signs are based on the changes of the system.
- q is negative when the system transfers heat to the surroundings.
- w is negative when the system does work on the surroundings.

17.2 The First Law of Thermodynamics

- **Objectives**
  - Define the first law of thermodynamics.
  - Describe how heat, work, and energy are related by the first law of thermodynamics.
  - Relate internal energy and enthalpy.

First Law of Thermodynamics

- **First law of thermodynamics** is the law of conservation of energy; energy can neither be created nor destroyed.
- **Internal energy, E**, is the total energy of the system and is a state function.

Internal Energy, E

- When a change occurs in a closed system (one in which energy, but not matter, exchanges with the surroundings), the change in internal energy, ΔE, is given by
  $\Delta E = q + w$
- ΔE is negative if the system transfers energy to the surroundings.

Measuring Internal Energy

- To evaluate ΔE:
  $\Delta E = q + w$
  $\Delta E = q - P\Delta V$
  At constant volume $\Delta V = 0$
  $\Delta E = q_v$
- ΔE is determined from constant-volume calorimetry.

Constant-Volume (Bomb) Calorimetry

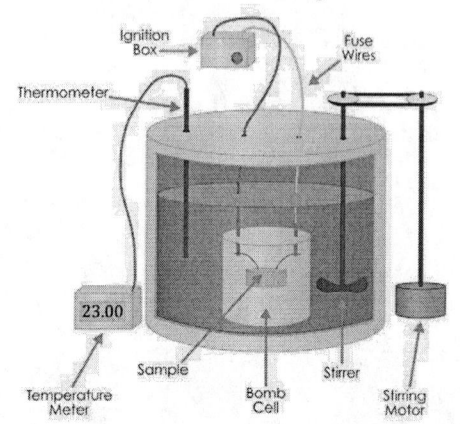

Example: Bomb Calorimetry

- The combustion of a 0.440 g sample of ethanol (C_2H_5OH, molar mass = 46.07 g/mol) in a bomb calorimeter with a heat capacity of 5.278 kJ/°C causes the temperature to rise from 23.98 °C to 26.45 °C. What is ΔE for the reaction?

  $C_2H_5OH\ (\ell) + 3\ O_2\ (g) \rightarrow 2\ CO_2\ (g) + 3\ H_2O\ (\ell)$

Energy and Enthalpy

- $\Delta E = q + w$
- At constant pressure,
  - ΔH is defined as the heat evolved, q_p, thus $\Delta E = q_p + w$.
- Substitute ΔH for q_p and $-P\Delta V$ for w.

  $\Delta E = \Delta H - P\Delta V$

  $\Delta H = \Delta E + P\Delta V$

PV Work

- $\Delta H = \Delta E + P\Delta V$
- The difference between change in enthalpy and internal energy is PV work and is significant only for reactions that involve gases.
- From the ideal gas law, $PV = nRT$, so
- $\Delta H = \Delta E + (\Delta n)RT$
- Δn = change in number of moles of gas

Example: Calculating ΔH

- Calculate ΔH at 25 °C and 1.00 atm pressure for the following reaction.

  $2NO_2 \text{ (g)} \rightarrow N_2O_4 \text{ (g)}$ $\Delta E = -54.7$ kJ

Test Your Skill: Calculating ΔH

- The products of the combustion of liquid glycerin, $C_3H_8O_3$, are gaseous carbon dioxide and liquid water. Balance the reaction for the combustion of 2 mol of glycerin and calculate ΔH for this reaction at 298K if burning a 1.240 g sample of glycerin in a bomb calorimeter with a heat capacity of 5.786 kJ/°C to increase in temperature from 23.45°C to 27.30°C.

17.3 Entropy and Spontaneity

- **Objectives**
  - Define entropy
  - Predict the sign of entropy changes for phase changes
  - Define and apply the second law of thermodynamics to chemical systems
  - Recognize that absolute entropies can be measured because the third law of thermodynamics defines a zero point

Disorder

- An increase in **disorder** (also known as randomness) is an important driving force for many changes.
- Two different gases, initially separated by a partition, will mix spontaneously when the partition is removed, increasing the disorder of the system.

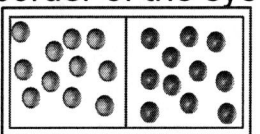

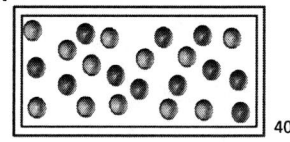

Entropy

- **Entropy (S)** is the thermodynamic state function that describes the amount of disorder.
- A large value for entropy means a high degree of disorder.

Entropy Change (ΔS)

- The entropy of a system generally increases when a solid dissolves in a liquid.

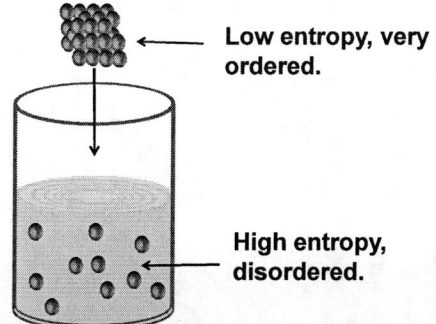

Low entropy, very ordered.

High entropy, disordered.

Entropy Change (ΔS)

- The entropy of a system generally decreases when a gas dissolves in a liquid.

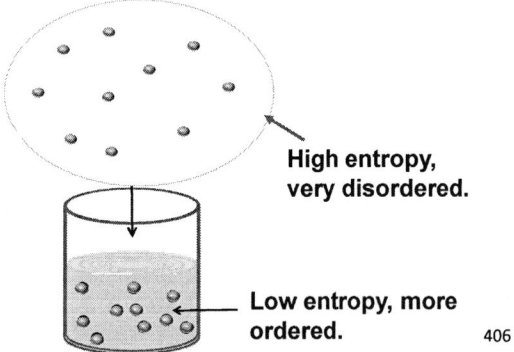

High entropy, very disordered.

Low entropy, more ordered.

Entropy Change (ΔS)

- An increase in disorder results in an increase in entropy. Some general guides are:
  - The entropy of a substance increases when solid becomes liquid, and when liquid becomes gas.
  - The entropy generally increases when a solute dissolves.
  - The entropy decreases when a gas dissolves in a solvent.
  - The entropy increases as temperature increases.

Example: Entropy Changes

- In the labelled boxes, draw atomic level images of the phases. For each phase change, label the arrows, and predict the sign of ΔS.

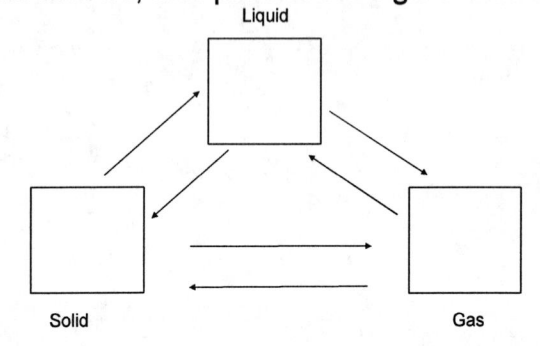

Liquid

Solid   Gas

Test Your Skill: Entropy Changes

Change	Sign of ΔS
Water freezes	
$Ba(OH)_2$ (s) + $2NH_4Cl$ (s) → $BaCl_2$ (aq) + $2NH_3$ (g) + $2H_2O$ (ℓ)	
CO_2 (g) + H_2O (ℓ) → H_2CO_3 (aq)	
Temperature of the planet increases	

Second Law of Thermodynamics

- The **second law of thermodynamics** states that in any spontaneous process, the entropy of the universe increases.

  $\Delta S_{univ} > 0$ for a spontaneous process.

Sign of ΔS_{univ}	Spontaneous Process
+	Forward
-	Reverse
0	Equilibrium

$$\Delta S_{univ} = \Delta S_{sys} + \Delta S_{surr}$$

- ΔS_{univ} provides no information about ΔS_{sys}.
- ΔS_{sys} is always negative for
  N_2 (g) + $3H_2$ (g) → $2NH_3$ (g)
  because the disorder of the system decreases.
- Under conditions when this reaction is spontaneous, ΔS_{surr} must be positive and larger than the negative ΔS_{sys}.

The Third Law of Thermodynamics

- The **third law of thermodynamics** states that the entropy of a perfect crystal of a substance at absolute zero is equal to zero (0).
- A perfect crystal at 0 K has no disorder.
- Unlike enthalpy and internal energy, absolute values of entropy can be determined.

Absolute Entropies

- In Appendix G, absolute entropies are given for substances in their standard state at 298 K.
- The entropy change for a reaction is given by:

$\Delta S°_{rxn} = \sum nS°[\text{prods}] - \sum nS°[\text{reacts}]$

where n is the coefficient of the products and reactants in the reaction.

Example: Entropy

- Calculate the standard entropy change for the reaction

$2C_2H_6(g) + 7O_2(g) \rightarrow 4CO_2(g) + 6H_2O(\ell)$

Substance	S°, J/ mol-K
$C_2H_6(g)$	229
$CO_2(g)$	214
$O_2(g)$	205
$H_2O(\ell)$	70

Test Your Skill: Entropy

- The combustion of ethane is spontaneous, but $\Delta S°$ is -617 J/K. Explain why this does not violate the second law of thermodynamics.

17.4 The Gibbs Free Energy

- **Objectives**
  - Define Gibbs free energy and relate the sign of a Gibbs free energy change to the direction of spontaneous reaction.
  - Predict the influence of temperature on Gibbs free energy.

Free Energy and ΔS_{univ}

- J.W. Gibbs defined a state function called the Gibbs free energy, G:

  $G = H - TS$

- At constant temperature and pressure, this becomes

  $\Delta G = \Delta H - T\Delta S$

- ΔG is directly related to ΔS_{univ}.

Free Energy and Spontaneity

- ΔG is a state function of the system.
- For any spontaneous change,
  - $\Delta S_{univ} > 0$
  - $\Delta G < 0$

Spontaneous reaction	ΔS_{univ}	ΔG
Forward reaction	Positive (+)	Negative (-)
At equilibrium	0	0
Reverse reaction	Negative (-)	Positive (+)

Standard Free Energy of Formation

- The standard free energy of formation is the free energy change to form one mole of a compound from its elements in their standard state.
- The standard state is:
  - Gas - 1 atm
  - Solutions - 1 M
  - Liquids and Solids – Pure
- Standard states are designated with a superscript: ΔG°, ΔH°, ΔS°

Calculating Standard Free Energy of a Reaction

- $\Delta G^\circ = \sum n\Delta G_f^\circ[\text{prods}] - \sum n\Delta G_f^\circ[\text{reacts}]$ can be used at 298 K.
- $\Delta G^\circ = \Delta H^\circ - T\Delta S^\circ$, can be calculated from tabulated data at any temperature because ΔH° and ΔS° are approximately constant as temperature changes.

Example: Calculate ΔG°

- Given the following chemical reaction and data at 298 K:

$$N_2(g) + 3H_2(g) \rightleftharpoons 2NH_3(g)$$

	N_2	H_2	NH_3
ΔG_f°, kJ/mol	0	0	-16
ΔH_f°, kJ/mol	0	0	-46
S°, J/mol-K	192	131	192

- Calculate ΔG° at 298 K and 1000 K. Assume ΔH° and ΔS° do not change with temperature.

ΔG and Spontaneity

- From the equation $\Delta G = \Delta H - T\Delta S$, a negative ΔH and a positive ΔS favor spontaneity.
- When the system is at equilibrium, $\Delta G = 0$.

Temperature Dependence of ΔG

- The dependence of ΔG on temperature arises mainly from the $T\Delta S$ term in the definition of ΔG.

$$\Delta G = \Delta H - T\Delta S$$

- ΔG is dominated by ΔH at low temperatures and by $-T\Delta S$ at high temperatures.
- Negative values of ΔH and positive values of ΔS favor spontaneity.

Direction of Spontaneous Reaction

$$\Delta G = \Delta H - T\Delta S$$

ΔH	ΔS	$-T\Delta S$	Temperature	ΔG	Spontaneous Direction
−	+	−	All	−	Forward
−	−	+	Low	−	Forward
			High	+	Reverse
+	+	−	Low	+	Reverse
			High	−	Forward
+	−	+	All	+	Reverse

ΔG is dominated by ΔH at low temperatures and by $-T\Delta S$ at high temperatures.

Example: Temperature Dependence

- Calculate the temperature for which $\Delta G°$ is 0 for the reaction below. Assume $\Delta H°$ and $\Delta S°$ do not change with temperature.

$$CS_2(\ell) \rightleftharpoons CS_2(g)$$

- At 298 K, $\Delta H° = 28$ kJ and $\Delta S° = 86$ J/K.
- Note: When the liquid and gas are at equilibrium, the temperature equals the boiling point of the substance.

17.5 Gibbs Free Energy and the Equilibrium Constant

- **Objectives**
  - Determine the effect of concentration on Gibbs free energy
  - Calculate the standard Gibbs free energy change from the equilibrium constant, and vice versa
  - Determine the effect of temperature on the equilibrium constant
  - Understand the connection between Gibbs free energy and work

Concentration and Free Energy

- Concentrations (or partial pressures) of reactants and products influence the free energy change of a reaction according to the equation

$$\Delta G = \Delta G° + RT \ln Q$$

where ΔG is the free energy change under non-standard conditions and Q is the reaction quotient (see chapter 14).

Example: Concentration Dependence

- For the reaction

$$2NO_2(g) \rightleftharpoons N_2O_4(g)$$

$\Delta G_f°$ kJ/mol   51   98

- Calculate $\Delta G°$ at 298 K.
- Calculate ΔG at 298 K when P_{NO_2} = 0.12 atm and $P_{N_2O_4}$ = 0.98 atm.

Free Energy and K_{eq}

- $\Delta G = \Delta G° + RT \ln Q$
- For a system at equilibrium, $\Delta G = 0$, and $Q = K_{eq}$, so
- $\Delta G° = -RT \ln K_{eq}$
- $\Delta G°$, calculated from values of $\Delta H°$ and $\Delta S°$, can be used to calculate the value of the equilibrium constant at any temperature.

Example: K_{eq} Calculation

- Given the following equation and the data below calculate the equilibrium constant (K_{eq}) at 298 K.

$$2NO(g) + Br_2(g) \rightleftharpoons 2NOBr(g)$$

ΔG_f°, kJ/mol	86.5	3	82
ΔH_f°, kJ/mol	90	31	82
S_f°, J/mol·K	211	245	274

Test Your Skill: K_{eq} Calculation

- Given the following equation and the data below calculate the equilibrium constant (K_{eq}) at 800 K.

$$2NO(g) + Br_2(g) \rightleftharpoons 2NOBr(g)$$

ΔG_f°, kJ/mol	86.5	3	82
ΔH_f°, kJ/mol	90	31	82
S_f°, J/mol·K	211	245	274

Temperature and K_{eq}

- The temperature dependence of K_{eq} is derived from two equations given earlier:
- $\Delta G^\circ = -RT \ln K_{eq} = \Delta H^\circ - T\Delta S^\circ$

$$\ln K_{eq} = \Delta S^\circ/R - \Delta H^\circ/RT$$

- The temperature dependence can also be expressed as:

$$\ln \frac{K_1}{K_2} = \frac{\Delta H^\circ}{R}\left(\frac{1}{T_2} - \frac{1}{T_1}\right)$$

Temperature and K_{eq}

$\ln K_{eq} = \Delta S°/R - \Delta H°/RT$

- A graph of $\ln K_{eq}$ vs. $1/T$ gives a straight line with slope $= -\Delta H°/R$ and an intercept of $\Delta S°/R$.

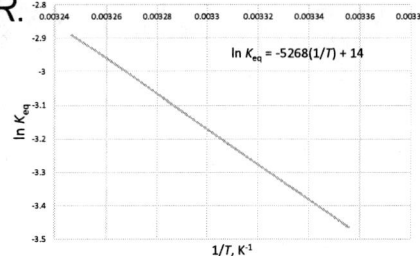

$\ln K_{eq} = -5268(1/T) + 14$

T, $\Delta G°$ and K_{eq}

- The temperature dependence of ΔG comes from the sign of ΔS.
    $\Delta G = \Delta H - T\Delta S$
- The temperature dependence of K_{eq} comes from the sign of ΔH.
    $\ln K_{eq} = \Delta S°/R - \Delta H°/RT$
  - If the reaction is exothermic, increasing T decreases the value of K_{eq}, as predicted by Le Chatelier's principle.

Example: Concentration and Temperature Dependence

- Calculate the partial pressure of $CS_2(g)$ at 310 K. Assume $\Delta H°$ and $\Delta S°$ are constant.
    $CS_2(\ell) \rightleftharpoons CS_2(g)$
- At 298 K, $\Delta H° = 28$ kJ and $\Delta S° = 86$ J/K.

Useful Work

- The change in free energy is the maximum work that can be performed by a spontaneous chemical reaction at constant temperature and pressure.
- $w_{max} = \Delta G$
- When $\Delta G > 0$ (spontaneous in the reverse direction), it represents the minimum work that must be provided to cause a change.

Chapter 18

Electrochemistry

18.1 Oxidation Numbers

- **Objectives**
  - Define oxidation, reduction, and redox reactions.
  - Assign oxidation numbers to atoms in chemical species.

Oxidation and Reduction

- **Oxidation** is the loss of electrons by a chemical process.
- **Reduction** is the gain of electrons by a chemical process.

Oxidation-Reduction ("Redox Reaction")

- An oxidation-reduction reaction, or redox reaction, is one in which electrons are transferred from one species to another.
- For example:

$$2Na(s) + Cl_2(g) \rightarrow 2NaCl(s)$$

- Sodium is oxidized to Na^+ and Cl_2 is reduced to 2 Cl^-.

Oxidizing and Reducing Agents

- The oxidizing agent is the reactant reduced; it gains electrons.
- The reducing agent is the reactant oxidized; it loses electrons.

Oxidizing and Reducing Agents

$$2Na(s) + Cl_2(g) \rightarrow 2NaCl(s)$$

- Sodium is oxidized to Na$^+$ and Cl$_2$ is reduced to 2 Cl$^-$.
- Sodium is the reducing agent and Cl$_2$ is the oxidizing agent.

Half-Reactions

- In a half-reaction, either the oxidation or the reduction part of a redox reaction is given, showing the loss or gain of the electron(s).

$$2Na(s) + Cl_2(g) \rightarrow 2NaCl(s)$$

Na $\rightarrow$ Na$^+$ + e$^-$ Oxidation half-reaction

Cl$_2$ + 2e$^-$ $\rightarrow$ 2 Cl$^-$ Reduction half-reaction

Oxidation States

- The oxidation state is
  - the charge on the monatomic ion or
  - the charge atoms would posses if the shared electrons are allocated to the more electronegative atom.
- Electron pairs shared by two atoms of the same element are divided equally.
- For the ionic compound CaCl$_2$:
  - Calcium has an oxidation state of +2
  - Chlorine has an oxidation state of -1

Assigning Oxidation Numbers

- Oxidation numbers for atoms in their elemental form are 0.
- The oxidation number of a monatomic ion is equal to the charge on the ion.
- In compounds, F is always -1. Other halogens are also -1 unless they are combined with a more electronegative element (i.e. O or a halogen above it on the periodic table).

Assigning Oxidation Numbers

- In compounds, O is -2 except for peroxides (-1) or when combined with F (+2).
- In compounds, H is +1 except in metal hydrides (-1)
- The sum of all the oxidation numbers of the atoms in a substance must equal the charge on the substance.

Example: Assigning Oxidation Numbers

- Assign oxidation numbers for each atom in $K_2Cr_2O_7$.

Test Your Skill: Oxidation Number

- Assign oxidation numbers to all elements on both sides of the reaction arrow for:

    $2SO_2 + O_2 \rightarrow 2SO_3$

Example: Redox Reactions

- From the previous Test Your Skill,
    - Determine which element is oxidized.
    - Determine which element is reduced.
    - Identify the oxidizing agent.
    - Identify the reducing agent.
    - How many electrons were transferred in the reaction as written?

18.2 Balancing Oxidation-Reduction Reactions

- **Objectives**
    - Balance oxidation-reduction reactions using the half-reaction method.
    - Balance redox reactions in both acidic and basic solutions.

Balancing Redox Equations

1. Determine oxidation numbers that change
2. Write the skeleton (unbalanced) half-reactions.
3. Balance element being oxidized or reduced.
4. Balance O by adding H_2O as needed.
5. Balance H by adding H^+ as needed.
6. Balance charges by adding e^- as needed.

Balancing Redox Equations

7. Multiply one or both half-reactions by an integer so that the number of electrons in both are the same.
8. Add the two half-reactions, canceling out the electrons and any components that appear on both sides of the equation.

Example: Balancing Redox Equations

- Balance the following equation.

$$Cr_2O_7^{2-} + C_2H_5OH \rightarrow Cr^{3+} + CO_2$$

Test Your Skill: Balancing Redox Equations

- Balance the following equation:

  $Cu + NO_3^- \rightarrow Cu^{2+} + NO$

Balancing Redox Equations in Basic Solutions

- In basic solutions, first balance in acid and then add OH^- to both sides of the reaction to eliminate any H^+.

  $H^+ + OH^- \rightarrow H_2O$

Example: Balancing Redox Reactions in Basic Solutions

- The following reaction is known to occur in basic solution. Balanced in acid:

  $3ClO^- + 2MnO_2 + H_2O \rightarrow 3Cl^- + 2MnO_4^- + 2H^+$

- Balance the reaction in base.

Test Your Skill: Balancing Redox Equations in Base

- Balance the following equation as a basic solution:

$$Zn + ClO^- \rightarrow Zn(OH)_4^{2-} + Cl^-$$

18.3 Voltaic Cells

- **Objectives**
  - Define and identify the components of a voltaic cell
  - Write half-cell reactions and the overall reaction from a diagram of a voltaic cell.
  - Identify the direction of flow of electrons and ions through a salt bridge in a voltaic cell.
  - Sketch half-cells that involve metal/metal ion, metal ion/metal ion, and gas/ion redox processes.

Voltaic Cells

- A **voltaic cell** or a galvanic cell is an apparatus that produces electrical energy directly from a redox reaction.

Voltaic Cell

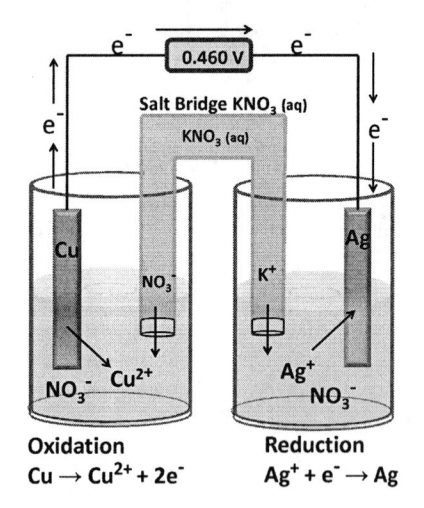

Overall Reaction and Shorthand Notation

- The right-hand cell contains the reduction by convention.
- The overall reaction is

   $Cu(s) + 2Ag^+(aq) \rightarrow Cu^{2+}(aq) + 2Ag(s)$

- The shorthand notation uses a single vertical line for a phase boundary and a double vertical line for the salt bridge.

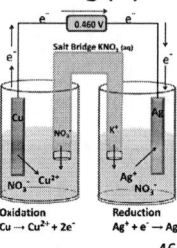

   $Cu \mid Cu^{2+} \parallel Ag^+ \mid Ag$

Voltaic Cells

- The oxidation half-cell contains the half-reaction:

   $Cu(s) \rightarrow Cu^{2+}(aq) + 2e^-$

- The reduction half-cell contains the half-reaction:

   $Ag^+(aq) + e^- \rightarrow Ag(s)$

- Since the half-cells are physically separated, the electrons must travel from one side to the other through a connecting wire.

Electrodes and Half-Cells

- Voltaic cells with redox reactions not involving neutral metals use an inert electrode, like gold, platinum or carbon, to provide electrical contact.
- Examples:
- Two soluble ions: $Fe^{3+}(aq) + e^- \rightarrow Fe^{2+}(aq)$
- Gas-ion reactions: $Cl_2(g) + 2e^- \rightarrow 2Cl^-(aq)$
- Insoluble salts: $AgCl(s) + e^- \rightarrow Ag(s) + Cl^-(aq)$

Inert-Electrode Half-Cell

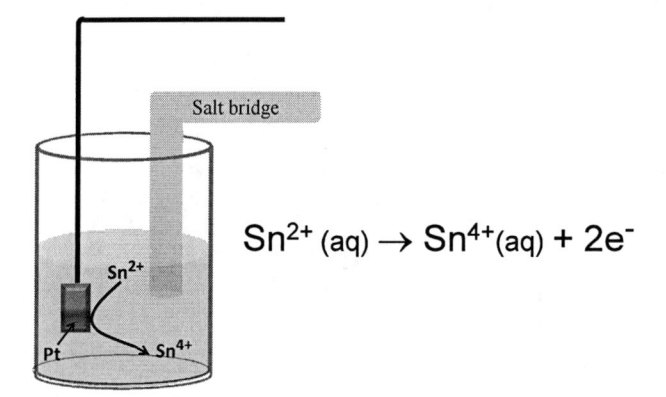

$Sn^{2+}(aq) \rightarrow Sn^{4+}(aq) + 2e^-$

18.4 Potentials of Voltaic Cells

- **Objectives**
  - Relate cell potential to a spontaneous reaction
  - Calculate the standard potential of a voltaic cell by combining two half-reactions
  - Use reduction potentials to predict the spontaneity of chemical reactions under standard conditions

Standard Potentials

- **Standard potential**, $E°_{cell}$, is the cell potential when each component in the reaction is present in its standard state.
- Solids, liquids and gases are in their pure state at 1 atm pressure
- Solutes have 1.0 *M* concentration
- The overall potential of a cell depends on the concentrations of the components in the reaction.

Potentials of Voltaic Cells

- For the reaction

    Cu(s) + 2Ag⁺(aq) → Cu²⁺(aq) + 2Ag(s)

    $E°_{cell}$ is +0.460 V.

- A positive cell potential means that the reaction is spontaneous in the forward direction.

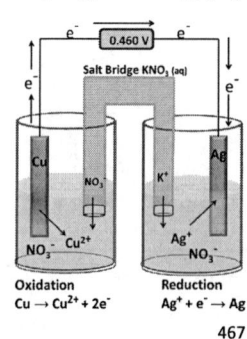

Cell Potentials

- Cell potentials are additive:

Cu(s) + 2Ag⁺(1 *M*) → 2Ag(s) + Cu²⁺(1 *M*)   +0.46 V
Zn(s) + Cu²⁺(1 *M*) → Cu(s) + Zn²⁺(1 *M*)   +1.10 V
Zn(s) + 2Ag⁺(1 *M*) → 2Ag(s) + Zn²⁺(1 *M*)   +1.56 V

Standard Reduction Potential

- The standard reduction potential for the reduction of H$^+$ is defined as 0 at 298 K and 1 atm.

    $2H^+ (aq, 1M) + 2e^- \rightarrow H_2 (g, 1\ atm)$

- The standard reduction potential of all other half-reactions is the potential of the reduction relative to the standard hydrogen electrode.

Standard Reduction Potential

- If the reduction reaction is reversed to make an oxidation reaction, change the sign of the potential.

    $Ag^+(aq, 1\ M) + e^- \rightarrow Ag\ (s)$ $E° = 0.80$ V
    $Ag(s) \rightarrow Ag^+(aq, 1\ M) + e^-$ $E° = -0.80$ V

Standard Reduction Potentials

Reduction Half-Reaction	$E°$ (V)
$F_2(g) + 2e^- \rightarrow 2F^-(aq)$	2.87
$Ag^+(aq) + e^- \rightarrow Ag(s)$	0.80
$Fe^{3+}(aq) + e^- \rightarrow Fe^{2+}(aq)$	0.77
$Sn^{4+}(aq) + 2e^- \rightarrow Sn^{2+}(aq)$	0.15
$2H^+(aq) + 2e^- \rightarrow H_2(g)$	0.00
$Co^{2+}(aq) + 2e^- \rightarrow Co(s)$	-0.28
$Fe^{2+}(aq) + 2e^- \rightarrow Fe(s)$	-0.44
$Zn^{2+}(aq) + 2e^- \rightarrow Zn(s)$	-0.76
$Mg^{2+}(aq) + 2e^- \rightarrow Mg(s)$	-2.37

Calculating Cell Potential

- The standard potential of a cell reaction is given by
$$E°_{cell} = E°_{red} + E°_{ox}$$
- Both half-reactions must transfer the same number of electrons.
- Multiplying the coefficients of a half-reaction to balance the electrons does NOT change the potential.

$Fe^{3+}(aq) + e^- \rightarrow Fe^{2+}(aq)$ $E° = 0.77$ V
$2Fe^{3+}(aq) + 2e^- \rightarrow 2Fe^{2+}(aq)$ $E° = 0.77$ V

Example: Calculating Cell Potentials

- Balance and calculate the standard cell potential of:

$Fe^{3+}(aq) + Cu(s) \rightarrow Cu^{2+}(aq) + Fe^{2+}(aq)$

- $Ag^+(aq) + e^- \rightarrow Ag(s)$ 0.80 V
- $Fe^{3+}(aq) + e^- \rightarrow Fe^{2+}(aq)$ 0.77 V
- $Cu^{2+}(aq) + 2e^- \rightarrow Cu(s)$ 0.34 V
- $2H^+(aq) + 2e^- \rightarrow H_2(g)$ 0.00 V
- $Fe^{2+}(aq) + 2e^- \rightarrow Fe(s)$ -0.44 V

Test Your Skill: Calculating Cell Potentials

- Write the balanced redox reaction, calculate the standard cell potential, and determine if the reaction is spontaneous:

$Cu \mid Cu^{2+} \parallel H^+ \mid H_2$

- $Ag^+(aq) + e^- \rightarrow Ag(s)$ 0.80 V
- $Fe^{3+}(aq) + e^- \rightarrow Fe^{2+}(aq)$ 0.77 V
- $Cu^{2+}(aq) + 2e^- \rightarrow Cu(s)$ 0.34 V
- $2H^+(aq) + 2e^- \rightarrow H_2(g)$ 0.00 V
- $Fe^{2+}(aq) + 2e^- \rightarrow Fe(s)$ -0.44 V

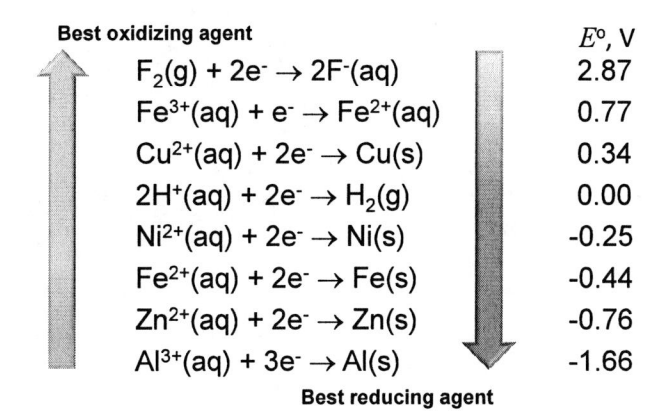

Activity Series

Best oxidizing agent

Reaction	$E°$, V
$F_2(g) + 2e^- \rightarrow 2F^-(aq)$	2.87
$Fe^{3+}(aq) + e^- \rightarrow Fe^{2+}(aq)$	0.77
$Cu^{2+}(aq) + 2e^- \rightarrow Cu(s)$	0.34
$2H^+(aq) + 2e^- \rightarrow H_2(g)$	0.00
$Ni^{2+}(aq) + 2e^- \rightarrow Ni(s)$	-0.25
$Fe^{2+}(aq) + 2e^- \rightarrow Fe(s)$	-0.44
$Zn^{2+}(aq) + 2e^- \rightarrow Zn(s)$	-0.76
$Al^{3+}(aq) + 3e^- \rightarrow Al(s)$	-1.66

Best reducing agent

Use of Activity Series

- Components with large positive reduction potentials are oxidizing agents - they oxidize components below them in the activity series.
- Components with large negative reduction potentials are reducing agents - they reduce components above them in the activity series.

Example: Activity Series

- Using the activity series below, determine what reaction, if any, occurs when Fe(s) is added to 1 M solutions of (a) $Zn(NO_3)_2$ and (b) $Co(NO_3)_2$.

$Co^{2+}(aq) + 2e^- \rightarrow Co(s)$
$Fe^{2+}(aq) + 2e^- \rightarrow Fe(s)$
$Zn^{2+}(aq) + 2e^- \rightarrow Zn(s)$

18.5 Cell Potentials, ΔG, and K_{eq}

- **Objectives**
  - Relate cell potential, free energy, and the equilibrium constant.
  - Calculate the equilibrium constant for a reaction from the standard potential of a voltaic cell.

Cell Potentials and ΔG

- $\Delta G = -nFE$ where:
  - ΔG = free energy change
  - n = number of electrons transferred
  - F = Faraday constant, 96,485 C/mol e⁻
  - E = cell potential

Cell Potentials and ΔG

- Under standard conditions,
  $$\Delta G° = -nFE°$$
- Positive cell potentials represent spontaneous reactions.

Example: Calculating $\Delta G°$

- Calculate $\Delta G°$ for:
$$AgCl(s) \rightarrow Ag^+(aq) + Cl^-(aq)$$
Given the following data:
$AgCl(s) + e^- \rightarrow Ag(s) + Cl^-(aq) \quad E° = 0.222 \text{ V}$
$Ag(s) \rightarrow Ag^+(aq) + e^- \quad E° = -0.80 \text{ V}$

Test Your Skill: Calculating $\Delta G°$

- Use the data below to calculate $\Delta G°$ for the reaction.
$$Al + Fe^{3+} \rightarrow Al^{3+} + Fe$$

$Fe^{3+} + 3e^- \rightarrow Fe \quad E° = -0.11 \text{ V}$
$Al^{3+} + 3e^- \rightarrow Al \quad E° = -1.66 \text{ V}$

Relating K_{eq} to $E°$

Electrochemistry: $\Delta G° = -nFE°$
Thermodynamics: $\Delta G° = -RT \ln K_{eq}$
$$-nFE° = -RT \ln K_{eq}$$
$$E° = \frac{RT}{nF} \ln K_{eq} = \frac{2.303 RT}{nF} \log K_{eq}$$
At 298 K, $E° = \frac{0.0591}{n} \log K_{eq}$
$$\log K_{eq} = nE°/0.0591$$
$$K_{eq} = 10^{nE°/0.0591}$$

Example: Calculating K_{eq} from E^o

- Determine K_{eq} for the following reaction at 298K

$$Al(s) + Fe^{3+}(aq) \rightarrow Al^{3+}(aq) + Fe(s)$$
$$E^o = +1.55 \text{ V}$$

Test Your Skill: Calculating K_{eq} from E^o

- Determine K_{eq} for the following reaction at 298 K.

$$Fe(s) + Pb^{2+}(aq) \rightarrow Pb(s) + Fe^{2+}(aq)$$
$$E^o = +0.31 \text{ V}$$

18.6 Dependence of Voltage on Concentration: The Nernst Equation

- **Objectives**
  - Use the Nernst equation to find the voltage of cells under nonstandard conditions of concentration
  - Calculate the concentration of ions in solution from measured cell potentials

The Nernst Equation

- The Nernst equation is used to calculate cell potentials under non-standard conditions.

$$E_{cell} = E°_{cell} - \frac{2.303RT}{nF} \log Q$$

at 298 K, $E_{cell} = E°_{cell} - \frac{0.0591}{n} \log Q$

where Q is the reaction quotient.

Example: The Nernst Equation

- Calculate the potential for the cell below, at 298 K. $E° = 1.10V$

Zn | Zn^{2+}(0.500 M) || Cu^{2+}(1.500 M) | Cu

Test Your Skill: The Nernst Equation

- Calculate the potential for the cell below, at 298 K. $E° = -0.78V$

Cu | Cu^{2+}(1.250 M) || Fe^{2+}(0.100 M) | Fe